T0074081

Cognitive Technologies

For further volumes:
http://www.springer.com/series/5216

Lutz Frommberger

Qualitative Spatial Abstraction in Reinforcement Learning

Dr.-Ing. Lutz Frommberger
Cognitive Systems Group
Department of Mathematics and Informatics
University of Bremen
P.O. Box 330 440
28334 Bremen
Germany
lutz@informatik.uni-bremen.de

This thesis was accepted as doctoral dissertation by the Department of Mathematics and Informatics, University of Bremen, under the title "Qualitative Spatial Abstraction for Reinforcement Learning". Based on this work the author was granted the academic degree Dr.-Ing.

Date of oral examination: 28th August 2009

Reviewers:

Prof. Christian Freksa, Ph.D. (University of Bremen, Germany)
Prof. Ramon López de Mántaras, Ph.D. (Artificial Intelligence Research Institute, CSIC, Barcelona, Spain)

Cognitive Technologies ISSN 1611-2482
ISBN 978-3-642-16589-4 e-ISBN 978-3-642-16590-0
DOI 10.1007/978-3-642-16590-0
Springer Heidelberg Dordrecht London New York

ACM Computing Classification: I.2

Cover design: KünkelLopka GmbH, Heidelberg

Printed on acid-free paper

Springer is part of Springer Science+Business Media (www.springer.com)

Foreword

Teaching and learning are difficult tasks not only when people are involved but also with regard to computer programs and machines: When the teaching/learning units are too small, we cannot express sufficient context to teach a differentiated lesson; when they are too large, the complexity of the learning task can increase dramatically such that it will take forever to teach and learn a lesson. Thus, the question arises, how we can teach and learn complex concepts and strategies, or more specifically: How can the lesson be structured and scaled such that efficient and effective learning can be achieved?

Reinforcement learning has developed as a successful learning approach for domains that are not fully understood and that are too complex to be described in closed form. However, reinforcement learning does not scale well to large and continuous problems; furthermore, knowledge acquired in one environment cannot be transferred to new environments. Although this latter phenomenon also has been observed in human learning situations to a certain extent, it is desirable to generalize suitable insights for application also in new situations.

In this book, Lutz Frommberger investigates whether deficiencies of reinforcement learning can be overcome by suitable abstraction methods. He discusses various forms of spatial abstraction, in particular qualitative abstraction, a form of representing knowledge that has been thoroughly investigated and successfully applied in spatial cognition research. With his approach, Lutz Frommberger exploits spatial structures and structural similarity to support the learning process by abstracting from less important features and stressing the essential ones. The author demonstrates his learning approach and the transferability of knowledge by having his system learn in a virtual robot simulation system and consequently transferring the acquired knowledge to a physical robot.

Lutz Frommberger's approach is influenced by findings from cognitive science. In this book, he focuses on the role of knowledge representation for the learning process: Not only is it important to consider *what* is represented, but also *how* it is represented. It is the appropriate representation of an agent's perception that enables generalization in the learning task and that allows for reusing learned policies in new contexts—without additional effort. Thus, the choice of spatial representation for the agent's state space is of critical importance; it must be well considered by the designer of the learning system. This book provides valuable help to support this design process.

Bremen, September 2010 *Christian Freksa*

Preface

Abstraction is one of the key capabilities of human cognition. It enables us to conceptualize the surrounding world, build categories, and derive reactions from these categories to cope with different situations. Complex and overly detailed circumstances can be reduced to much simpler concepts, and not until then does it become feasible to deliberate about conclusions to draw and actions to take.

Such capabilities, which come easily to a human being, can still be a big challenge for an artificial agent: In the past years of research I investigated how to employ such human concepts in a learning machine. In particular, my research focused on utilizing spatial abstraction techniques in agent control, using the machine learning paradigm of reinforcement learning. This led to results published in journals and conference proceedings over the years that are now integrated and significantly extended to a comprehensive study on spatial abstraction in reinforcement learning in this book. It is spans the whole range from formal aspects to empirical results.

Reinforcement learning allows us to learn successful strategies in domains that are too complex to be described in a closed model or in cases where the system dynamics are only partially known. It has been shown to be effectively applicable to a large number of tasks and applications. However, reinforcement learning in its "pure" form shows severe limitations in practical use. In particular, it does not scale well to large and continuous problems, and it does not allow for reuse of already gained knowledge within the learning task or in new tasks in unknown environments. Spatial abstraction is an appropriate way to tackle these problems.

When regarding the nature of abstraction, I believe that only a consistent formalization of abstraction allows for a thorough investigation of its properties and effects. Thus, I present formal definitions that distinguish between three different facets of abstraction: aspectualization, coarsening, and conceptual classification. Based on these definitions it can be shown that aspectualization and coarsening can be utilized to achieve the same effect. Hence, the process of aspectualization is to be preferred when using spatial abstraction in agent control processes, as it is computationally simple and its features are easily accessible. This allows for coping even with high-dimensional state spaces. The property of a representation being aspectualizable turns out to be central for agent control.

In order to use abstraction to control artificial agents, I argue for an action-centered view on abstraction that concentrates on the decisions being drawn at certain states. I derive criteria for efficient abstraction in agent control tasks and show that these criteria can most satisfactorily be matched by the use of qualitative representations, especially when they model important aspects in the state space such that they can be accessed by aspectualization.

In sequential decision problems we can distinguish between goal-directed and generally sensible behavior. The corresponding spatial features form task space and structure space. As it is of special importance to describe structural elements of the state space explicitly in an abstract spatial representation, I introduce the concept of structure space aspectualizable observation spaces. For this kind of state space, two methods are developed in this book: task space tile coding (TSTC) and a posteriori structure space transfer (APSST). They allow for reusing structural knowledge while learning to solve a task and also in different tasks in unknown environments. Furthermore, I introduce structure-induced task space aspectualization (SITSA), a mechanism for situation-dependent spatial abstraction based on knowledge gained from a structural analysis of learned policies in previous tasks.

We will study the effect of the proposed techniques on an instance of structure space aspectualizable state spaces, namely le-RLPR, an abstract spatial representation tailored for robot navigation in indoor environments. It describes the circular order of landmarks around the moving robot and the relative position of walls with regard to the agent's moving direction. Compared to coordinate-based metrical approaches, le-RLPR enables us to learn successful strategies for goal-directed navigation tasks considerably faster. Policies learned with le-RLPR also allow for generalization within the actual learning task as well as for transferring knowledge to new scenarios in unknown environments. As a final demonstration we will see that RLPR-based policies learned in a simulator can also be transferred to a real robotics system with little effort and allow for sensible navigation behavior of a robot in office environments.

Acknowledgments

At this point I want to express my gratitude to several people who helped me during my work on this book.

First of all, I thank Christian Freksa for advising my doctoral thesis and giving me the opportunity to work in the Cognitive Systems research group at the University of Bremen. He brings together people from various scientific fields for interdisciplinary research. This provides an inspiring and productive atmosphere, and I am thankful that I was involved there for so many years. Christian has been always available when I needed advice. Often, his comments and ideas made me look at my work from a different point of view and thus broadened my mind.

Furthermore, I want to express my gratitude to Ramon López de Mántaras for his willingness to be a reviewer of my doctoral thesis and especially for his enthusiasm and his detailed and encouraging remarks on my work.

Particularly in the early stages of my research it had been important to receive encouraging feedback on the ideas I had. In particular, I thank Reinhard Moratz for initially supporting my approach. Furthermore, I thank Joachim Hertzberg, Frank Kirchner, Martin Lauer, George Konidaris, and Stefan Wölfl for inspiring and encouraging discussions that helped me to focus my work. Also, several anonymous reviewers provided substantial feedback on papers emerging from ongoing work on this book that I submitted to workshops, conferences, and journals.

Martin Riedmiller sparked my interest in reinforcement learning when I was a student at the University of Karlsruhe. I thank him for repeatedly giving me the opportunity to extensively discuss my work with him and his Neuroinformatics group at the University of Osnabrück. I acknowledge especially Stephan Timmer's valuable comments and hints regarding my approach.

I notably enjoyed working with my colleagues at the Cognitive Systems group, who gave me lots of feedback over the years. Especially, the graduate seminar was a great opportunity for inspiring discussions. I thank Diedrich Wolter for constantly pushing me forward and his help in making the nasty robot move. Also, I thank Mehul Bhatt, Frank Dylla, Julia Gantenberg, Kai-Florian Richter, Jan Frederik Sima, and Jan Oliver Wallgrün for volunteering to proofread parts of this book. I also thank my student co-workers for their dedication: Fabian Sobotka provided valuable assistance on the implementation of the software and Jae Hee Lee assisted in mathematical formalizations.

Money is not everything, but when available, it helps a lot. I thank the German Research Foundation (DFG) for its financial support of the R3-[Q-Shape] project of the Transregional Collaborative Research Center SFB/TR 8 Spatial Cognition, within which this work was carried out.

Most importantly, I thank my family, Michaela, Mara, and Laila. For many months I dedicated much of my time to writing this book rather than to them. I am deeply grateful for their support, their patience, and their love, without which finalizing this book would have been impossible.

Bremen, September 2010 *Lutz Frommberger*

Contents

Symbols

$\mathcal{A}$	Action space, 10
c_i	Sampled color view, 104
cert	Decision certainty for a state, 89
Conf	Decision confidence for a policy, 85
conf	Decision confidence for a state, 85
$\mathcal{D}$	Arbitrary domain, 45
E	Expected value (in statistics), 13
e	Eligibility trace, 18
H	Horizon, 11
h	Hash function, 76
$\mathcal{I}$	Initiation set for options, 31
I^{κ}	Aspectualization index vector, 52
I^{κ}	Inverse aspectualization index vector, 52
L^*	Set of all detected landmarks, 100
L_i	Sector for landmark selection, 100
$l_{\max}$	Maximum number of landmarks allowed in a sector, 101
$\mathcal{O}$	Observation space, 15
$\mathcal{O}_{\text{NDesc}}$	Set of non-decision structures, 90
$\mathcal{O}_S$	Structure space, 71
$\mathcal{O}_T$	Task space, 71
$p^{\max}$	Vector of maximum feature values, 79
$p_i^{\max}$	Maximum feature value of dimension i, 79
Q	Action-value function, 14
Q^*	Optimal action-value function, 19
Q_S	Structure space Q-function, 83
Q_T	Task space Q-function, 87
q_{π^*}	π^*-preservation quota, 56
$\mathcal{R}$	RLPR grid, 110
R	Reward function, 12
$\mathcal{S}$	State space, 10
s	State, 10

ssd	Structure space descriptor, 72
T	State transition function, 12
tsd	Task space descriptor, 72
V	Value function, 13
V^*	Optimal value function, 13
α	Learning rate, 18
β_i	Landmark sample angle, 104
χ	Tiling function, 76
γ	Discount factor, 13
δ	Temporal difference error, 18
ε	Exploration probability, 17
ζ	Motion noise parameter, 124
Θ	SITSA abstraction function, 90
κ	Abstraction function, 45
λ	Trace decay parameter, 19
μ	Task/structure space weight, 87
ξ	SMDP waiting time parameter, 31
π	Policy, 11
π^*	Optimal policy, 11
π_S	Structure space policy, 72
ρ	Line detection distortion parameter, 140
σ	Landmark noise parameter, 139
τ	RLPR overlap status, 111
$\overline{\tau}$	RLPR overlap status (whole scene), 111
$\overline{\overline{\tau}}$	RLPR overlap status (whole scene, boolean), 111
τ'	RLPR adjacency status, 111
$\overline{\tau}'$	RLPR adjacency status (whole scene), 111
$\overline{\overline{\tau}}'$	RLPR adjacency status (whole scene, boolean), 111
ψ	Observation function, 15
ψ_S	Structure space observation function, 114
ψ_T	Task space observation function, 100
ω	Observation in an observation space, 15
$\overline{\omega}$	Structure space observation, 83
$\perp$	Empty symbol, 90
$\equiv_2$	modulo 2, 15
$\sim$	Similarity operator, 55

Acronyms

A-CMAC	Averager cerebellar model articulator controller
APSST	A posteriori structure space transfer
CMAC	Cerebellar model articulator controller
MDP	Markov decision process
QSR	Qualitative spatial reasoning
POMDP	Partially observable Markov decision process
RL	Reinforcement learning
RLPR	Relative line position representation
SDALS	Structural-decision-aware landmark selection
SITSA	Structure-induced task space aspectualization
SMDP	Semi-Markov decision process
TSTC	Task space tile coding

Chapter 1
Introduction

One of the most essential properties of a cognitive being is its ability to learn. Learning is the "process of acquiring modifications in existing knowledge, skills, habits, or tendencies through experience, practice, or exercise" (Encyclopædia Britannica, 2007). These modifications lead to a performance improvement of the cognitive being (also called *cognitive agent*) in the tasks it has to solve in its daily routines. Learning provides the agent with a preferably good adaptation of its behavior to the situations it is confronted with.

While most of the learning efforts of human beings and animals are achieved in the early years, learning is generally a life-long process. Perceived situations may change over time, and even the perception abilities themselves may change, and the dynamics of the agent may vary due to age or abrasion. These changes require continuous adaptations of the acquired strategies and behaviors over a longer period of time. It is desirable to find this ability also in artificial cognitive agents, for example, in autonomous robots.

1.1 Learning Machines

Much effort has been spent in the field of artificial intelligence (AI) to investigate methods for machine learning. This field of research spans two distinct paradigms. *Supervised learning* requires external knowledge given by an expert, who supervises the learning process. The learning agent is supposed to find a mapping between its input values and the desired output that is given by the expert. In contrast, *unsupervised learning* autonomously constructs a classification of the input without intervention from outside. Many types of machine learning approaches exist somewhere between supervised and unsupervised learning.

This book concentrates on one of the most influential machine learning techniques: the learning paradigm of *reinforcement learning* (RL) (Sutton and Barto, 1998). In RL, learning does not take place by teaching or supervision, but by interaction with a dynamic and uncertain environment. It can be seen as a form of

L. Frommberger, *Qualitative Spatial Abstraction in Reinforcement Learning*, Cognitive Technologies, DOI 10.1007/978-3-642-16590-0_1,

weakly supervised learning. The concept of reinforcement learning was addressed very early in psychology and cybernetics and has gained a still increasing popularity in machine learning research over the last two decades. Basically, it implements a mechanism of reinforcing tendencies that lead the system to a "positive" state. Reinforcement learning is trial-and-error learning. Positive reinforcement is only given when the system reaches a well-defined goal state. The aim of this mechanism is to find the optimal way to reach this goal state. This way is given by a sequence of actions, each usually performed after a decision at a given, discrete point in time. Reinforcement learning methods are mostly applied to operate on a special case of sequential decision problems, so-called *Markov decision processes* (MDPs).

Reinforcement learning is very valuable when the characteristics of the underlying system are not known and/or difficult to describe or when the environment of an acting agent is only partially known or completely unknown. Various applications have been realized with reinforcement learning approaches, mostly concerning game playing, robotics, and control problems.

1.1.1 An Agent Control Task

Autonomous agents are in continuous interaction with the world they are operating in. Navigation in space, which is an essential ability of such agents, is a complicated process of perceiving the environment with their sensory system and performing physical actions according to an adequate interpretation of the collected sensory data. What is adequate in this context depends on the problem the agent has to solve.

Example 1.1. Imagine a discrete grid world with 6×6 grid cells (Fig. 1.1). An agent is always within one of the grid cells and can go from there to the adjacent grid cells in cardinal directions. The world is unknown to the agent, and its task is to reach a specified goal location from any position within the grid. There are 36 different positions the robot can be in, the *system states* or, for short, the *states*. This problem

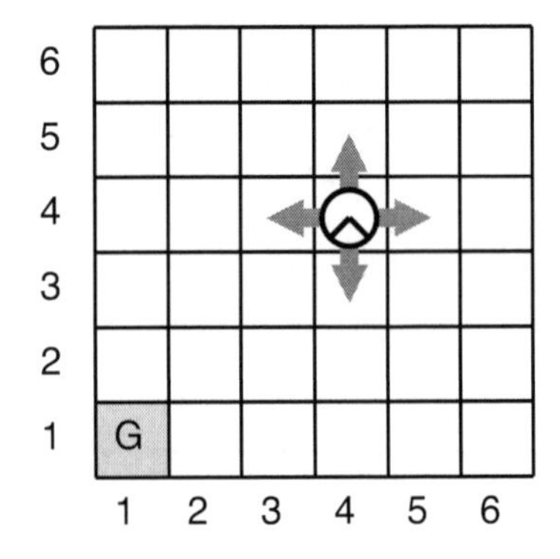

Fig. 1.1 A grid cell example. The robot is at position $(4,4)$. Its goal (G) is to reach the bottom left cell $(1,1)$. Its primitive actions are movements to neighbored cells to the left, right, top, and bottom of its position

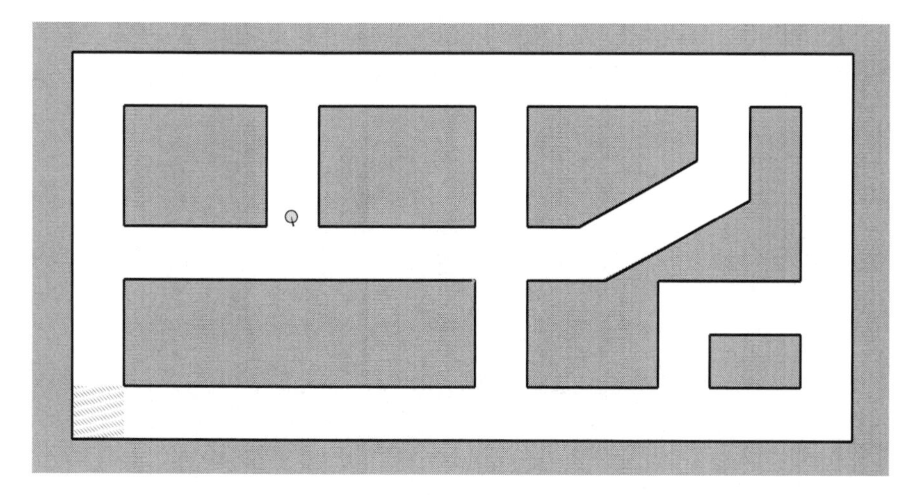

Fig. 1.2: A robot in a simulated office environment: the state space of this problem is continuous

can be formulated as a Markov decision process and can therefore be solved with reinforcement learning.

During the training process, the agent learns a *policy* that returns a particular action to execute for every state the agent is in. The policy is based on the *value function* that maintains an assessment of states with regard to solving the overall problem. For the simple problem in Example 1.1, a reinforcement learning algorithm is able to learn a solution after executing a few hundred actions. The complexity of RL scales linearly with the number of states: To give an impression, Kaelbling et al. (1996) report the need for 531,000 learning steps for a grid world with 3,277 states.

Long training times are a general problem of reinforcement learning. RL methods are proved to converge to an optimal solution, but the prerequisite is that *each* system state be continuously updated—which is practically impossible in larger state spaces. Even worse, most real-world state spaces are not discrete. Figure 1.2 shows a robot in an office environment—a continuous world with an infinite number of states.

An important question that arises here is how to describe a system state in a given context. Sensor readings of the robot are given in real numbers and form a continuous state space. That means that the value function has a continuous domain and cannot be stored easily in a table as in the case of a discrete one. To cope with continuous state spaces, some kind of *value function approximation* is used. Various approaches exist for this. What is common to all of them is that the incorporation of these methods introduces, besides a bunch of new parameters to cope with, uncertainty in the representation that may have unwanted effects. If the approximation is too rough, states may not be distinguished even if they had to be; if it is too fine, the training times will become unacceptably long. The choice of the right function approximation and the choice of its parameters usually requires solid expert knowledge by the designer of the learning system. In large and continuous environments reinforcement learning becomes cumbersome.

1.1.2 Structure of a State Space

The agent's behavior within its environment is based on the structure the surrounding world offers. In an office building, for example, the world is structured into rooms, corridors, and open spaces, and these structural elements induce certain actions of the agent.

In a seminal paper, Thrun and Schwartz (1995) claim that for being able to adapt reinforcement learning to more complex tasks it is necessary to discover the structure of the world and abstract from its details. Consequently, recent approaches take the structure of the world into account when trying to cope with the problems named above. Various ways of addressing this issue have been proposed: One is working on the action level, for instance, adapting the actions of the agents and collapsing them into larger sequences. Another idea is to regard the way the value function is stored by introducing value function approximation based on the structure and geometry of the given problem.

In contrast to that, the approach proposed in this book is to explicitly model the structural elements of the world within the state space representation. The underlying idea is that structurally similar parts of the state space should also have a similar representation so that the agent's view on the problem is generally the same when the structure of the environment is more or less identical. To create such representations, *abstraction* is a key issue. This abstraction is applied directly to the original state space, which is defined by the agent's sensory system, such that the learning process itself operates on abstract concepts instead of metrical measurements. In this case, we speak of *spatial abstraction.*

1.1.3 Abstraction

Abstraction is a reduction of data that omits information that is not necessary. This book proposes distinguishing three different facets of abstraction: From an original feature vector, the number of features can be lowered by omitting some of them (aspectualization), the granularity of the features can be reduced (coarsening), or completely new features can be built based on combinations of the original ones (conceptual classification). Especially, aspectualization is an interesting concept, because it allows for easy access to subsets of a feature vector. We call a representation that allows for aspectualization an *aspectualizable* representation, and we will see that they are particularly essential for successful spatial abstraction in agent control tasks.

Abstraction is not a blind destruction of available information—it aims at achieving a classification that is appropriate for a specialized task. In the good old days of early AI research, the aim was to gather all knowledge in one big formal representation. Nowadays, this view on the problem has changed. Sloman (1985) argues that specialized formalisms for different purposes are a more promising approach to building intelligent systems. Following that, this book argues for choosing abstraction methods according to the task at hand. The goal is to show that a clever choice of abstraction techniques allows for efficient and stable solutions for the given problems. This book shows that the the success of learning heavily depends on the chosen state space representation.

To get such a state space description, we use the paradigm of *qualitative abstraction* to create a spatial representation of the world. The resulting *qualitative spatial representation* neglects all the metrical details the sensors offer and builds classes of spatial concepts instead. Humans use those kinds of concepts as well: Things are "to my left" or "to my right," and most of the time exact angles and distances do not matter. Distinctions between classes are made exactly when they invoke a difference in the choice of action to take. For example, it is not necessarily bad to head toward an abyss—but having reached the abyss, it is time to react and stop moving forward. Similarly, the autonomous robot in Fig. 1.2 does not need care about its precise distance from walls when following a corridor unless this distance gets too small. The change from one abstract concept to another describes structural elements of the underlying world and, following Thrun and Schwartz (1995), this structure is crucial in the success of the learning algorithm. Qualitative representations of space, when designed properly, make this structure explicit.

1.1.4 Knowledge Reuse

By applying qualitative spatial abstraction, improvements in learning speed and robustness can be achieved. This alone would certainly be a satisfactory justification for its use. However, the chosen abstraction paradigm offers additional benefits that go far beyond that kind of performance tuning. The explicit modeling of structural elements can also enable the agent to reuse gathered knowledge learned at one location in the environment that is structurally similar to its current location. For example, if a child has once learned to follow a corridor, it does not have to learn this again in every new corridor. This ability to apply a learned strategy to unknown instances within the same task is called *generalization*. This work shows that the use of qualitative abstraction can lead to a generalizing behavior of the agent almost without any further effort, which again leads to a significant improvement in the learning performance.

While generalization describes a property within the same task, it is also desirable that a learning agent can reuse a learned behavior or at least parts of it in a completely different task, for example, in an unknown environment with a different goal to reach. For humans, this is a natural property: For instance, if you once have learned to succeed in playing a strategy game, you might show an easier learning of other strategy games—because you have achieved an understanding of this type of problem. But in general, this is not the case with reinforcement learning approaches as described here, because a completely new set of state-action pairs with different action selections is necessary to succeed in the new task. The question of whether and how a learned solution can be reused in partially similar settings and thus speed up the learning process is an increasingly important issue in current machine learning research. Recently the term *transfer learning* became popular to describe this research field. In contrast to generalization, which describes the ability to apply a learned strategy to unknown instances within the same task, transfer learning tackles generalization ability across different tasks. In a way, transfer learning can be seen as "cross-task-generalization" from a source task to a target task.

There must be things that are common to both the source and the target tasks so that transfer of learned strategies makes sense. The claim of this book is that those commonalities are of a structural sort. Therefore, if it is possible to identify these common structural elements and also to isolate them within the learned strategy, then the transfer problem scales down dramatically. For example, each office environment is structured by walls. Thus, the position of walls should be explicated in the state space representation to facilitate knowledge transfer.

This work shows that qualitative spatial representations are perfectly suitable to explicitly model such structures, in particular, if the representation is chosen to be *aspectualizable*. Then, the idea is that if the spatial representation is (partly) the same in different tasks, there is no need to transfer at all, and learned strategies can be used in the target task as they are. Qualitative spatial abstraction makes it possible to scale down the transfer problem to the choice of the state space representation.

All in all, this book aims at moving the focus to the question of how to represent state spaces. As the approaches presented in this work exclusively operates on the level of post-processing sensory information, they can easily be combined with any appropriate existing learning methodology. This provides an expressive means at hand to use machine learning approaches in a profitable way: faster and more robust learning, less parameters, intuitive usage, easy communicability, combinability, and knowledge transfer properties—everything just due to the choice of a qualitative spatial representation that is in line with the given problem.

1.2 Thesis and Contributions

This is the thesis of this book:

> Aspectualizable qualitative spatial abstraction is the key to enable reinforcement learning to efficiently solve tasks in complex and continuous state spaces.

In particular, this kind of abstraction

- provides manageable state spaces that improve learning performance significantly and
- allows for reuse of learned knowledge within the same task as well as in completely unknown environments.

Regarding these claims, this work makes the following main contributions:

- This book provides a formal theory of abstraction, defining the abstraction facets aspectualization, coarsening, and conceptual classification. Aspectualization is identified as a powerful means to represent knowledge, and aspectualizability is shown to be a critical property of a representation for knowledge reuse. Consequentially, qualitative spatial representations are best suited to explicating the structure of a state space and fulfilling the requirements of a suitable abstraction in agent control tasks.
- It is shown in this work that an agent's behavior in control tasks can be divided into two aspects. We distinguish between goal-directed and generally sensible behaviors that form the task space and the structure space. Based on that, the concept of structure space aspectualizable state spaces is defined and shown to be the key to generalization and knowledge transfer. Algorithms for generalization and transfer learning are defined to benefit us from previously learned knowledge within the same or completely new tasks.
- Tailored for robot navigation in indoor environments, the structure space aspectualizable spatial representation le-RLPR is developed. It is shown that this representation leads to a significant improvement in learning performance and allows for reusing learned policies within or across tasks and spatial environments. In particular it is described how the use of le-RLPR can even bridge the gap between simplified simulation and real-world robotics.

1.3 Outline of the Thesis

The remainder of this book is organized as follows: In Chap. 2, the learning paradigm of reinforcement learning and the Markov framework are introduced along with basic notations. Readers familiar with reinforcement learning can easily skip this chapter.

Remaining challenges in reinforcement learning are identified in Chap. 3, and the state of the art in application of reinforcement learning to complex and continuous state spaces and knowledge transfer is worked out and analyzed.

Chapter 4 introduces a formal theory of abstraction and investigates interdependencies of its different aspects. Criteria for efficient spatial abstraction in agent control tasks are defined. Especially, qualitative spatial representations are highlighted.

Chapter 5 is dedicated to the questions of generalization and transfer learning. There, we take a thorough look at the relation of state space representation and knowledge transfer and describe formalisms to achieve policies that also operate in different environments than the one learned in.

A spatial representation tailored for indoor robot navigation tasks is introduced in Chap. 6. The role of landmarks and structural information is discussed, and the proposed representation is analyzed with regard to its properties and suitability according to the requirements worked out in the previous chapters.

Chapter 7 is dedicated to an empirical evaluation of the proposed methods and representations with regard to performance in reinforcement learning tasks. It especially demonstrates how qualitative spatial abstraction can be used for knowledge transfer from a simplified simulation to a real-world robotics task.

Finally, Chap. 8 summarizes this book and provides an outlook on further research following the work presented here.

Chapter 2
Foundations of Reinforcement Learning

This chapter gives an introduction to the machine learning paradigm of reinforcement learning and introduces basic notations. Following a short overview on machine learning in Sect. 2.1, Sect. 2.2 explains the reinforcement learning model, before the central framework of Markov decision processes is described in Sect. 2.3. The importance of exploration is explained in Sect. 2.4, and temporal difference learning, including Q-learning, is presented in Sect. 2.5. The chapter closes with an overview on performance measures (Sect. 2.6).

Readers familiar with reinforcement learning can safely skip this chapter and proceed with Chap. 3, which identifies central challenges in reinforcement learning that are addressed within this book.

2.1 Machine Learning

Machine learning is one of the broadest subfields of artificial intelligence. To exactly define what machine learning is about we refer to a definition by Nilsson (1996), who states that "a machine learns whenever it changes its structure program or data based on its inputs or in response to external information in such a manner that its expected future performance improves." That means machine learning is concerned with the desire to build algorithms that enable a computer not only to fulfill a given task according to predefined rules, but also to adapt and improve its behavior over time while interacting with the environment it is operating in.

Roughly, machine learning can be divided into three cases: supervised learning, unsupervised learning, and reinforcement learning.

1. *Supervised learning* describes the process of acquiring a mapping function from given input values to predefined desired output values. This is achieved by giving examples of input-output pairs. In the case of a discrete output space, this process is called *classification*; in a continuous output space we speak of *regression*. An example of supervised learning can be found in speech recognition, where

L. Frommberger, *Qualitative Spatial Abstraction in Reinforcement Learning*, Cognitive Technologies, DOI 10.1007/978-3-642-16590-0_2,

phonetic data samples have to be mapped to certain utterings; usually, Hidden Markov Models (HMMs) are used for this purpose. Other techniques to achieve such a mapping in supervised learning include neural networks, case-based reasoning, support vector machines, and many more. For an overview see Russell and Norvig (2003).

2. *Unsupervised learning* is not based on output given a priori. The goal is to autonomously find patterns in a given data set. The most common examples for unsupervised learning are reduction of dimensionality and clustering, which can be achieved by statistical methods, for example, principal component analysis. An unsupervised learning process can be seen as a sophisticated form of data processing.
3. *Reinforcement learning* can be seen as somewhere between supervised and unsupervised learning. The learning system receives a feedback for its actions, but it is not provided with examples to learn from. According to Russell and Norvig (2003), it is the most general of the three cases, because it does not rely on a teacher but learns, as the name says, on the basis of received reinforcement. It is this learning paradigm that is exclusively investigated within this book.

2.2 The Reinforcement Learning Model

An important difference between reinforcement learning and both supervised and unsupervised learning is that in the beginning of the reinforcement learning process no data is available at all, so the learning literally starts from scratch. In a reinforcement learning problem as described by Kaelbling et al. (1996) a so-called *agent* operates in an environment and gathers data from constant interaction with its surroundings. From this interaction, the agent receives a *reinforcement*. In combination with the interaction patterns, the reinforcement provides a basis for the learning process. Another, more frequently used term for reinforcement is *reward*.

As the agent starts to learn from scratch, an improvement must be achieved by trial-and-error. The agent performs actions—which due to missing knowledge are random in the beginning—and adapts its behavior based on experienced rewards. This constitutes a control loop of the *reinforcement learning model*. As defined by Kaelbling et al. (1996), this model consists of

- a set of environment states $\mathcal{S}$, the *state space*,
- a set $\mathcal{A}$ of actions the agent can perform, the *action space*,
- a set of reinforcement signals, usually $\mathbb{R}$ or a subset of it.

$\mathcal{S}$ and $\mathcal{A}$ define the *domain* of the problem.

The corresponding control loop is depicted in Fig. 2.1: The system is in a state $s \in \mathcal{S}$. Based on s and a reinforcement signal r the agent chooses an action $a \in \mathcal{A}$ which changes the system state s, and the cycle starts again.

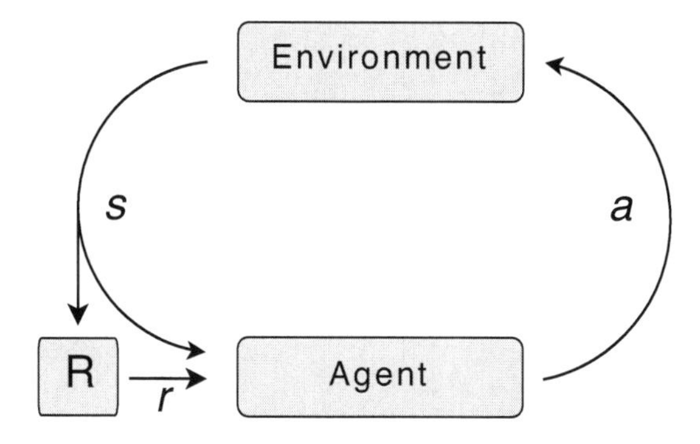

Fig. 2.1 The simple reinforcement learning model. The agent executes an action a, which has an effect on the environment. Its state s is perceived by the agent, coupled with a reinforcement signal r that depends on the environmental state. Based on s and r, a new action is chosen

Both $\mathcal{S}$ and $\mathcal{A}$ can be discrete or continuous sets. Continuous sets are to be expected in manifold real-life control tasks; for example, both a plane's flying altitude (state) and a change in its engine pressure (action) are encoded as real numbers.

In the reinforcement learning problem actions are executed as a sequence of decisions at discrete points in time t. Thus, these problems are labeled *sequential decision problems*. In a state $s_t \in \mathcal{S}$ at a time point t, the agent chooses an action $a_t \in \mathcal{A}$ and receives a reward r_t.

The solution to the reinforcement learning problem is the determination of which action to take for any system state. This is an optimization task: The agent has to develop a *policy* π which maps an action to a system state—so π is a function $\pi : \mathcal{S} \rightarrow \mathcal{A}$. A policy that does not change over time is called a *stationary policy*. The desired *optimal policy* π^* is the policy that maximizes the rewards achieved over time. Thus, in contrast to supervised learning, there are no "correct" input/output-pairs—which action will be the best in a certain state depends on the overall performance of the agent and evolves over time within the learning process.

Reinforcement learning takes place in repetitive cycles which we call *episodes*. An episode begins with the agent being in an *initial state* $s_S \in \mathcal{S}$. It repeatedly chooses an action $a \in \mathcal{A}$ which brings it to the next state and delivers a reinforcement signal r. An episode ends when the agent reaches a *terminal state*. Another possibility to end an episode is if a previously defined number of steps H, the so-called *horizon*, has been executed without reaching a terminal state.

2.3 Markov Decision Processes

The action selection in a state s is not trivial, because the goal is to choose the action that maximizes the received reward over the whole sequence of actions and not just the reinforcement given in a particular state. In a way, the agent has to look into the future, as the reward feedback that influences its decision may occur several time steps ahead. This is known as the concept of *delayed rewards*. Sutton and Barto (1998) name delayed rewards along with trial-and-error search as one of the two main characteristics of reinforcement learning. A concept to adequately express stochastic model behavior and delayed rewards is the *Markov decision process* (MDP) (Bellman, 1957).

2.3.1 Definition of a Markov Decision Process

A Markov decision process is defined as a tuple $\langle \mathcal{S}, \mathcal{A}, T, R \rangle$. It is a special case of the reinforcement learning model described in Sect. 2.2. In detail, it consists of the following:

- A set of states $\mathcal{S}$.
- A set of actions $\mathcal{A}$.
- $T : \mathcal{S} \times \mathcal{A} \times \mathcal{S} \to [0,1]$, a state transition function. $T(s,a,s')$ delivers the probability that applying a certain action $a \in \mathcal{A}$ in a state $s \in \mathcal{S}$ will bring the system to state $s' \in \mathcal{S}$. This maps the usually non-deterministic system behavior. Assuming that s_t is the current state at a time point t and s_{t+1} is the subsequent state after the execution of an action a_t, $T(s,a,s') = P(s_{t+1} = s' | s_t = s, a_t = a)$. In the case of a deterministic underlying system, $T(s,a,s')$ will be either 0 or 1, so $T : \mathcal{S} \times \mathcal{A} \times \mathcal{S} \to \{0,1\}$ will hold.[1]
- $R : \mathcal{S} \times \mathcal{A} \to \mathbb{R}$, a reward function that assigns an *immediate reward* $R(s,a)$ for action $a \in \mathcal{A}$ in state $s \in \mathcal{S}$.

Subsequent states in an MDP are fully described by the probability distribution given by T, so they depend only on the current state and the executed action. In particular, the next state does not depend on the way the current state was reached—past states or actions are irrelevant. This property is called the *Markov property* or the *Markov assumption*. If a system shows the Markov property, we often say that the system is "Markovian" or just "Markov."

Example 2.1. Let us again look at the grid world in Fig. 1.1 described in Example 1.1, with the agent in a discrete 6×6 grid with four possible actions to reach adjacent cells. The robot cannot leave the grid, so applying action a in state s results in a state

$$f_T(s,a) = (\min(6, \max(1, s_1 + a_1)), \min(6, \max(1, s_2 + a_2))).$$

Let us say that the agent receives a reward of 50 reaching the goal state and a reward of -1 for not. Formalized as an MDP this setting looks as follows

[1] It is also possible to formulate T as a function mapping a state-action pair to a probability distribution over the state space: $T : \mathcal{S} \times \mathcal{A} \to \Pi(\mathcal{S})$.

$$\begin{aligned}
\mathcal{S} &= \{s = (s_1, s_2) \in \mathbb{Z}^2 | s_1, s_2 \in [1,6]\}, \\
\mathcal{A} &= \{(0,1), (1,0), (0,-1), (-1,0)\} \subseteq \mathbb{Z}^2, \\
T(s,a,s') &= \begin{cases} 1 & s' = f_T(s,a) \\ 0 & \text{else} \end{cases}, \\
R(s,a) &= \begin{cases} 50 & \text{if } s = (1,1) \\ -1 & \text{else} \end{cases}.
\end{aligned}$$

Because $T(s,a,s') \in \{0,1\}$ for all $s, s' \in \mathcal{S}$ and $a \in \mathcal{A}$, this MDP is deterministic.

If the state space $\mathcal{S}$ of an MDP is not enumerable, which is, for example, the case for $\mathcal{S} = \mathbb{R}^n$, we speak of a continuous state space and a *continuous state MDP*. In the following, though, we assume that $\mathcal{S}$ is discrete.

2.3.2 Solving a Markov Decision Processes

For finding a solution to an MDP, that is, the optimal policy π^*, the key issue is the *expectation* of future rewards to come. This is coupled with the question of how many of the future rewards are to be taken into account and how this is handled. The most straightforward method is to optimize the sum of rewards r_t over a horizon of the next H time steps. However, the most popular model is the so-called *infinite horizon discounted model*. In this model, all following rewards are summed up, but are weighted by a *discount factor* $\gamma \in \mathbb{R}$ $(0 < \gamma < 1)$, so that future rewards have a smaller impact on the overall result. The function $V : \mathcal{S} \to \mathbb{R}$ assigns to every system state $s \in \mathcal{S}$ the *overall expected reward* when starting in s, the *value* of s. $V : \mathcal{S} \to \mathbb{R}$ is called the *value function* and in the infinite horizon discounted model is defined as

$$V(s) = E(\sum_{t=0}^{\infty} \gamma^t r_t) \tag{2.1}$$

with E denoting the expected value. The discount keeps the infinite sum bounded.

The infinite horizon discounted model gained its popularity because it has sound mathematical properties. For instance, in this model we have a guaranteed existence of an optimal deterministic stationary policy (Bellman, 1957), which is very convenient when applying machine learning methods. Thus, we concentrate on the infinite horizon discounted model in this work.

The optimal value of a state is the expected reward when always following the optimal policy. The optimal value function V^* is

$$V^*(s) = \max_{\pi} E(\sum_{t=0}^{\infty} \gamma^t r_t). \tag{2.2}$$

The value of the current state s is the immediate reward plus the value of the subsequent state s', so we get the optimal value by the following recursive equation considering the transition probabilities. This equation is known as the *Bellman optimality equation*:

$$V^*(s) = \max_{a \in \mathcal{A}}(R(s,a) + \gamma \sum_{s' \in \mathcal{S}} T(s,a,s')V^*(s')). \tag{2.3}$$

Having the optimal value function V^*, the optimal policy is to always execute the action returning the highest value:

$$\pi^*(s) = \operatorname{argmax}_{a \in \mathcal{A}}(R(s,a) + \gamma \sum_{s' \in \mathcal{S}} T(s,a,s')V^*(s')). \tag{2.4}$$

This type of action selection is called a *greedy policy*.

2.3.2.1 Value Iteration

An algorithm to learn the optimal value function that defines the optimal policy is the *value iteration* algorithm. The idea of value iteration is to iterate over all actions in all states and adapt the value function according to (2.3):

$$Q(s,a) = R(s,a) + \gamma \sum_{s' \in \mathcal{S}} T(s,a,s')V(s'). \tag{2.5}$$

$$V(s) = \max_{a \in \mathcal{A}} Q(s,a). \tag{2.6}$$

The *action-value function* $Q : \mathcal{S} \times \mathcal{A} \rightarrow \mathbb{R}$ denotes the overall expected reward when in state s and executing action a, so the value of s is the maximum of the Q-values. The given loop has to be executed until the emerging policy fulfills the expectations and can be considered "sufficiently good." Generally, it is not clear when this will be the case, even if some stopping criteria for the value iteration algorithm could be derived. However, it has been proved that value iteration converges to the optimal value function V^* (Bellman, 1957; Bertsekas, 1987). For a discussion of the complexity properties of this algorithm see Littmann et al. (1995).

The update procedure in (2.5) needs to have information on all the subsequent states s' that could possibly be reached. This is called a *full backup*. In real-world applications, full backups are usually impossible to achieve, unless a complete *model* of the system exists, that is, unless the exact consequences and possible subsequent states of an action are known. If there is no model and the agent has to detect the successor state by trying out a certain action—one of the key ideas of reinforcement learning—only knowledge about a single successor state can be gathered. This is called a *sample backup*. Value iteration also works with sample backups, as long as it can be ensured that all state-action pairs are visited sufficiently often. Repeated visits of the same state-action pair (s,a) then sample the probability distribution $T(s,a,s')$.

2.3.2.2 Policy Iteration

Another possible way to solve an MDP is by iterating over policies instead of states and actions; this is called *policy iteration*. An example of a policy-based learning method is the *adaptive heuristic critic* (AHC) algorithm (Barto et al., 1983). Policy iteration takes fewer iterations compared to value iterations, but the latter proves to be much faster in practice (Kaelbling et al., 1996). Thus, value-based methods are preferred over policy iteration when solving planning problems (LaValle, 2006). Policy iteration is not investigated in greater detail in this book.

2.3.3 Partially Observable Markov Decision Processes

The MDP model assumes that the current system state can always be determined completely. Furthermore, it assumes that this determination is correct. For real-world systems, this assumption rarely holds. An agent usually receives only a subset of the complete system state through its sensory system. This can be expressed with the help of an extension of the MDP model.

A *partially observable Markov decision process* (POMDP) (Lovejoy, 1991) takes into account that state spaces may not be fully observable; this means that the agent does not know its system state completely. This problem has been first investigated in the context of operations research (OR), but has gained great importance in the field of AI and machine learning (Cassandra et al., 1994; Littman et al., 1994).

A POMDP is defined as a tuple $\langle \mathcal{S}, \mathcal{A}, \mathcal{O}, T, R \rangle$, with $\mathcal{S}$, $\mathcal{A}$, T and R defined as in a normal MDP (Sect. 2.3). Additionally, there is

- the *observation space* $\mathcal{O}$, a set of possible *observations*, and
- an *observation function* $\psi : \mathcal{S} \to \mathcal{O}$ that delivers an observation $\omega \in \mathcal{O}$ for every system state $\mathcal{S}$. If this is non-deterministic, we have $\psi : \mathcal{S} \to \Pi(\mathcal{O})$, where $\Pi(\mathcal{O})$ is a probability distribution over $\mathcal{O}$ which represents the conditional probability $P(\omega|s)$.

An observation $\omega \in \mathcal{O}$ can be interpreted as the agent's perception of the environment.

Example 2.2. Let us consider the setting from Example 2.1. Now we assume that the agent can perceive the world at a granularity of only two cells. Without loss of generality (w.l.o.g.) it can handle even cell numbers only. This results in

$$\mathcal{O} = \{s = (s_1, s_2) \in \mathbb{Z}^2 | s_1, s_2 \in [1,6], s_1 \equiv_2 0 \wedge s_2 \equiv_2 0\},$$
$$\psi(s) = (2\lfloor((s_1+2)/2)\rfloor - 2,\ 2\lfloor((s_2+2)/2)\rfloor - 2).$$

The symbol $\equiv_2$ stands for "modulo 2." System states $(1,3)$ and $(0,2)$ share the same observation, as $\psi((1,3)) = (0,2) = \psi((0,2))$.

When taking the observation space $\mathcal{O}$ instead of $\mathcal{S}$, the resulting system no longer has the Markov property, in general. The underlying MDP is Markovian, of course;

the POMDP is not. However, POMDPs are Markovian relative to the so-called *belief state* (Aström, 1965). A belief state $b(s)$ denotes a probability distribution over $\mathcal{S}$ and represents the agent's belief in being in state s. If you take into account the set of belief states $\mathcal{B}$ instead of $\mathcal{S}$, the resulting model is Markovian again (Bertsekas, 1987) and results in a continuous state space MDP. Belief states can be updated considering an observation ω by using the Bayesian rule, and so POMDPs can be converted into MDPs. Hence, theoretically POMDPs can be solved with algorithms such as value iteration. Also, the Baum-Welch algorithm (Baum and Sell, 1968) can solve POMDPs. However, for larger domains the learning gets increasingly harder (Theocharous et al., 2005). For an overview on exact solution techniques refer to Kaelbling et al. (1998).

In general, solving POMDPs is a hard and—for larger state spaces—open issue. The problem of finding a policy for a POMDP is NP-hard (Littman, 1994). Retrieving exact solutions usually is infeasible for non-trivial POMDP state spaces, as explicit representations of believe states become intractable. According to Hasinoff (2003), who provides an excellent overview on POMDPs, most solution approaches to POMDPs rely on approximation techniques and heuristics or sophisticated mechanisms like memory-based algorithms or hierarchical architectures.

2.4 Exploration

Using sample backups to solve an MDP requires that all state-action pairs be visited sufficiently often. When following the greedy policy described above, this cannot be ensured. Most probably state-action pairs remain that are not visited at all, because their Q-value is always lower than that of others.

To make the agent visit those state-action pairs, there must be exceptions to the greedy action selection. This exceptional behavior is called *exploration*. A good exploration strategy has to make sure that the state-action space is explored sufficiently well. The opposite, acting according to the behavior induced by π, is called *exploitation*.

Clearly, there is a trade-off between exploration and exploitation. Exploration is needed to fully observe the state space, and exploitation is needed to benefit from the already learned policy. The more exploration takes place, the less exploitation can be done, and vice versa. But for successful learning, both are necessary. This antagonism is usually referred to as the *exploration-exploitation dilemma*. Some techniques to tackle this dilemma in reinforcement learning are given in Wilson (1996); and numerous further strategies are still being developed. For a taxonomy of exploration techniques refer to Thrun (1992).

2.4.1 ε-Greedy Action Selection

A simple but effective exploration method is an *ε-greedy* policy. This means that, given an *exploration probability* $\varepsilon \in [0,1]$, at each time step the agent follows the policy π with a probability of $1-\varepsilon$. In the other cases, a random action is executed.

ε-greedy policies are limited as they provide only one exploratory step with a probability of ε. For two exploratory steps in a row, the probability is ε^2, so it happens very rarely. ε-greedy policies keep the agent near the trajectories that are induced by π, and so they modify the learned behavior only slightly. Strategies that would require a larger step away from π may remain undiscovered. On the other hand, keeping the agent near a policy which already shows reasonable success may be promising for further improvement. One way to soften this problem is to start with a high value of ε, as in the beginning there is no reasonable policy to exploit at all. During learning, the value of ε is gradually diminished, so the role of exploration reduces as the quality of the policy increases. However, the choice of the parameter for diminishing is as crucial as the choice of ε itself.

2.4.2 Other Exploration Methods

An alternative to ε-greedy exploration is the so-called *R*-MAX approach (Brafman and Tennenholtz, 2003). There, the value function is initialized with numbers that exceed the maximum overall reward possible. Thus, each value function update will lead to state values lower than initialization. Under greedy action selection, the highest reward expectation now comes from previously unvisited states, such that the system prefers to visit states that have never been seen before.

To achieve a smart and comprehensive exploration of the state space, exploration is often guided by heuristics (see, for example, Zhao et al. (1999) or Bianchi et al. (2007)). Heuristic approaches are manifold, and a description would go beyond the scope of this chapter.

2.5 Temporal Difference Learning

When trying to adapt the values of states the problem is that it is not clear in advance whether an action taken was a good or a bad move. The overall consequences of this action cannot be determined immediately as rewards received in the future may have great impact. This problem is known as the *temporal credit assignment problem.*

This problem has been addressed with the introduction of *temporal difference learning*, or *TD learning* (Sutton, 1988). The idea is to update the value of state s *after* one or more actions have been executed, based on current estimates of the following state values. Put differently, the algorithm makes use of the knowledge gain during the time between the visit of s and the later update of $V(s)$. TD methods use sample backups.

2.5.1 *TD(0)*

The update rule of the so-called *TD(0)* method works similarly to that of value iteration. $V(s)$ is updated after the execution of an action a based on the immediate reward $r = R(s,a)$ and the current (discounted) value estimate of the subsequent state $V(s')$:

$$V(s) = \alpha\delta = \alpha(r + \gamma V(s') - V(s)). \tag{2.7}$$

The *learning rate* or *step size* α ($\alpha \in \mathbb{R}$) controls the speed of the update of $V(s)$. We call the term $\delta = r + \gamma V(s') - V(s)$ the *temporal difference error*. TD(0) has been proved to converge to the optimal value function (Sutton, 1988; Dayan, 1992).

TD methods are *model-free* methods. This means that for learning a policy no model of the underlying dynamic system is required: Updates are based on sampled experience and trial-and-error; knowledge of the implication of actions or the transition distribution is not necessary a priori. This in an important property of TD learning and one of the reasons for its success.

Examples for TD learning algorithms are *actor-critic methods* (Barto et al., 1983) and *SARSA* (Rummery and Niranjan, 1994; Sutton, 1996). However, this work focuses on the most popular of the TD methods, Q-learning, which is described in Sect. 2.5.3.

2.5.2 *Eligibility Traces/TD(λ)*

Updates of V as given in (2.7) are called *1-step backups*, as updates are only applied to the previous state value $V(s)$ based on δ. All other states that have been visited before s will not get updated, despite having contributed to reaching the state s' the agent is actually in. So for a state s we only regard rewards that will be given one time step in the future. Analogously, backups that regard more than one future state are called 2-step, 3-step, or generally n-step backups. A TD learning mechanism dealing with n-step backups is consequently called an *n-step TD method*. So we store an *eligibility trace* $e(s) \in \mathbb{R}^+$ for each state s that denotes its contribution to the actual state. At each step, in state s', all eligibility traces are updated as follows

$$e(s) = \begin{cases} \gamma\lambda e(s) + 1 & \text{if } s = s' \\ \gamma\lambda e(s) & \text{else} \end{cases}. \tag{2.8}$$

The parameter $\lambda \in [0,1]$, which gives the name to TD(λ) learning, is the *trace decay parameter*. At each time step, all eligibility traces are multiplied by λ, and the eligibility trace of the current state is incremented by 1, so earlier states receive less influence than later ones. For $\lambda = 0$, we have the 1-step backups of TD(0). A value of $\lambda = 1$ leads to no decay at all, so an update influences $V(s)$ identically for all previously visited states s, but this is practically never used.

The value update including eligibility traces is as follows

$$V(s) = \alpha \delta e(s) = \alpha e(s)(r + \gamma V(s') - V(s)). \quad \forall s \in \mathcal{S} \tag{2.9}$$

The update has to be performed for every single state. However, $e(s) = 0$ for almost all states, so updates in a real application only have to consider the limited number of recently visited states.

If an eligibility trace $e(s)$ is updated as described in (2.8), then it is called *accumulating* trace, because the values add up when visiting s again. Instead of that, traces can also be *replacing* (Singh and Sutton, 1996). Then, the update function is as follows

$$e(s) = \begin{cases} 1 & \text{if } s = s' \\ \gamma \lambda e(s) & \text{else} \end{cases}. \tag{2.10}$$

Using replacing traces prevents $e(s)$ from growing too large when s is repeatedly visited. This variant performs significantly better than accumulating traces.

2.5.3 Q-Learning

Q-learning (Watkins, 1989; Watkins and Dayan, 1992) is the most popular of the TD methods. It directly operates on state-action pairs and gets its name from the action-value function introduced in (2.5). Learning follows the update rule

$$Q(s,a) = Q(s,a) + \alpha(R(s,a) + \gamma \max_{a' \in \mathcal{A}} Q(s',a') - Q(s,a)) \tag{2.11}$$

with $s \in \mathcal{S}$ being the actual state, s' the following state, and $a \in \mathcal{A}$ an action.

Q-learning is labeled an *off-policy* algorithm, because it does not need the policy at all to update Q, which directly approximates the optimal action-value function Q^*. The policy determines which state-action pairs are sampled over the learning process. This simplicity made Q-learning attractive to the machine learning community and made it the learning algorithm used by a majority of reinforcement learning approaches in research and applications. Also, Q-learning is mathematically tractable and allowed for quite some convergence proofs and analyses. For example, it can be shown that Q-learning guarantees convergence towards Q^* (Watkins and Dayan, 1992; Tsitsiklis, 1994). Consequently, Sutton and Barto (1998) name Q-learning "one of the most important breakthroughs in reinforcement learning." The Q-learning algorithm is sketched in Algorithm 1.

Algorithm 1 The Q-learning algorithm

choose a starting position s
repeat
  choose $a = \text{argmax}_{a'} Q(s,a')$
  execute action a to reach s'
  retrieve reward $R(s,a)$
  update $Q(s,a)$ according to (2.11)
  $s \leftarrow s'$
until s is a terminal state

Q-learning can also be coupled with eligibility traces. The n-step version of Q-learning is called Q(λ). The mechanism described by Watkins in his Ph.D. thesis (Watkins, 1989) applies the ideas given in Sect. 2.5.2 to the Q-learning algorithm. Special attention has to be paid to exploration: It has to be ensured that eligibility is only set for sequences of state-action pairs that follow the greedy policy, because (suboptimal) exploratory actions falsify the contribution of the policy to the overall reward. So if an exploratory action is taken, all traces $e(s,a)$ have to be set to 0. This, of course, diminishes the effect of eligibility traces to some extent. Tackling this problem, Peng's Q(λ) (Peng and Williams, 1994, 1996) provides an alternative algorithm without the need for resetting traces. Although the use of this method no longer results in convergence to Q^*, the algorithm performs significantly better than Watkins's Q(λ) in empirical studies (Sutton and Barto, 1998). However, in this book we stick to Watkins's variant, as its implications are easier to understand and analyze, and absolute performance is not the main point of interest of the work presented here.

2.6 Performance Measures

To evaluate the success of a learning process, it is helpful to define some quality criteria. Kaelbling et al. (1996) name three performance measures, for all of which they identify certain pitfalls:

Eventual convergence to the optimal. Convergence to an optimal solution can be proved under certain circumstances. This is a nice property, but practically useless, because a fast convergence to a near-optimal solution may be more suitable in praxis.

Speed of convergence to optimality. An important question is how fast the learning algorithm finds the optimal policy π^*. But as optimality is only reached asymptotically, this measure is ill-defined according to Kaelbling et al. (1996).

Regret. To evaluate the performance of the system during learning, *regret* (Berry and Fristedt, 1985) defines the loss of performance of the actual system compared to the case in which it had been executing the optimal strategy from the beginning. However, results on regret are usually quite difficult to obtain.

None of the given measures is appropriate for evaluating the learning process in a practical context. Therefore, let us define some other criteria that are useful for judging the success for the methods introduced in this work:

Success rate. In goal-oriented tasks which we examine within this work it can clearly be determined whether this goal has been reached or not. Regardless of the optimality of the solution, the success rate defines the percentage of test runs that can solve the given task. Usually, we want to see a success rate near 100%. Less successful policies have to be dropped.

Speed to reach (near-)100% success. The number of learning episodes until the agent is able to fulfill the task in all or almost all cases regardless of optimality of solution is of course a critical performance measure.

Optimality of solution. If the achieved solution is not (yet) optimal, the deviation from the optimal solution after a given learning time is an important measure for judging the quality of the learned policy.

Speed to reach near-optimality. Because speed of convergence to optimality is ill-defined (see above), it is more suitable to define a region of near-optimality and regard the time in which the system provides such a near-optimal solution.

Robustness of the learning process. Learning processes practically never show monotonous improvement, so the performance can temporarily get weaker again while learning. We speak of *robust* learning if the success rate and optimality of solution do not show unstable development. Robust learning facilitates successful usage of learned policies after a short time, while non-robust learning can produce policies that perform unexpectedly badly over the learning process.

Forbidden states during learning. Because of the practical problems with regret, it is better to investigate how often the agent enters a “forbidden” state. These states could be ones that cause physical damage to the agent or require intervention by a human supervisor. The less often forbidden states are entered, the better. This measure is especially important when working with physical agents.

Number of parameters involved. While autonomous learning aims at finding a solution in overly complex tasks, the complexity of the design of the learning environment can prevent finding such a solution. Choosing the right parameters for the learning procedure does not rarely require solid expert knowledge. A small number of parameters to adjust increases the probability to find a good configuration. Generally, only having few parameters is recommended.

All these performance criteria can not be optimized in parallel. For example, more complex algorithms may yield better performance, but introduce additional parameters, or the solution will be closer to optimal at the cost of longer learning times. All of the given performance measures have their application in certain scenarios, and which one to focus on will generally depend on the task at hand. However, all of the given measures are criteria under which we evaluate the methods presented later in this work.

Chapter 3
Abstraction and Knowledge Transfer in Reinforcement Learning

This chapter presents the state of the art in research on reinforcement learning with a focus on abstraction and transfer learning. Especially, the open questions of performance in large, continuous state spaces and knowledge transfer are worked out as central challenges of reinforcement learning with regard to applications in Sect. 3.1. In the following, three approaches to tackle these problems are investigated: value function approximation (Sect. 3.2), temporal abstraction (Sect. 3.3), and spatial abstraction (Sect. 3.4). After the research field of transfer learning is introduced in Sect. 3.5, all the presented methods are discussed in Sect. 3.6, and guidelines for the approach described in this work are carried out there.

3.1 Challenges in Reinforcement Learning

Reinforcement learning has proved to be a valuable technique to solve sequential decision problems (Sect. 2.2), in particular, problems formulated within the Markov framework (Sect. 2.3). RL is able to achieve successful strategies in domains that are too complex to be described in a closed model and in cases where the system dynamics are only partially known. It has been shown to be effectively applicable to a large number of tasks and applications (Kaelbling et al., 1996). In particular, robot navigation has been a popular field of research in which RL has been frequently applied (see Smart and Kaelbling, 2000, for example).

However, reinforcement learning in its "pure" form shows severe limitations in practical use. In particular, we want to investigate two important issues here: First, the applicability of RL to complex state spaces as we encounter them in real-world problems, and second, the value and reusability of the gained knowledge for the current learning process—especially in a context of prolonged learning.

L. Frommberger, *Qualitative Spatial Abstraction in Reinforcement Learning*, Cognitive Technologies, DOI 10.1007/978-3-642-16590-0_3,

3.1.1 Reinforcement Learning in Complex State Spaces

One problem we encounter is that reinforcement learning does not scale well to larger problems. It is well known that RL as a method of dynamic programming suffers from the so-called *curse of dimensionality* (Bellman, 1957). This means that the size of the state space grows exponentially with the number of state variables. Although reinforcement learning techniques have been shown to be successful in solving even large problems and outperform other variants of dynamic programming (Sutton and Barto, 1998, p. 107), the resulting training times of RL often are unbearably long, because exploration of the state space needs an impractically large amount of time. In particular, a large number of training runs is needed, which makes unmodified reinforcement learning unsuitable for real-world problems that demand training on physically existing hardware such as autonomous robots.

Another problem relates to the internal representation of the value function. Storage within a lookup table becomes infeasible with regard to memory consumption in almost all cases except very small environments (Kaelbling et al., 1996). Especially crucial are scenarios where the domain $\mathcal{S}$ or the set of actions available $\mathcal{A}$ (or both) are continuous (see Smart and Kaelbling, 2000, for example). But these kinds of problems are encountered when solving real-world tasks. For continuous state spaces it is impossible to represent all system states as the states cannot be enumerated. As a consequence, approximating representations are needed to store the value function.

A further property of the real world is its partial observability. Reinforcement learning proves to be successful in solving Markov decision processes (MDPs), but the challenge of how to cope with partially observable Markov decision processes (POMDPs) with reasonable effort in large state spaces is still open (see Sect. 2.3.3). However, the real world must be seen as partially observable: For the learning machine, the perception of the world is provided by sensors, and no sensor can represent the whole world including all its facets. Also, it is impossible to collect all information that might be relevant to the system behavior.

It is widely accepted that for solving more complex tasks techniques have to be introduced into the learning process that abstract from the rich details of the state information, even if this results in suboptimal solutions.

3.1.2 Use and Reuse of Knowledge Gained by Reinforcement Learning

A crucial question is how to represent and utilize information already gained during the learning process. Solving an MDP means finding one singular optimal solution for one well-defined problem, specified by the given domain (that is, state space $\mathcal{S}$, action space $\mathcal{A}$, and transition probabilities T) and the reward function R, which specifies the goal. In other words, the learned policy π is applicable to the MDP

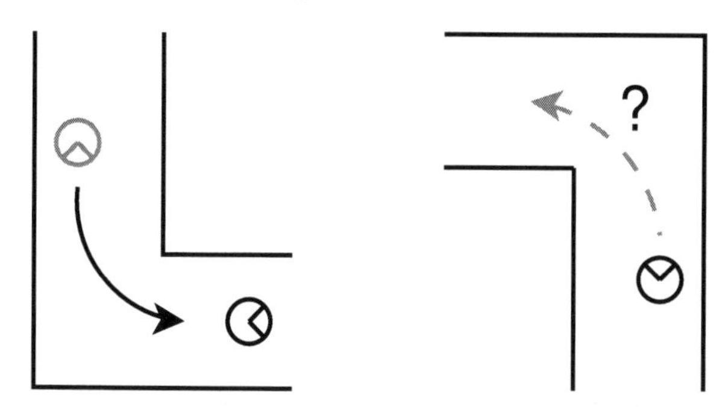

Fig. 3.1: Similar situations: two corridor bends with different spatial orientation. Will the agent be able to benefit from knowledge acquired in the left situation when trying to master the right situation? If coordinates and orientation are used to represent the state space, this is not the case

it is based on, but not applicable to other decision problems: Action selection for every state is taken with respect to the specific task to be solved; but no *general* knowledge about the task can be derived from the policy. In other words, the action acquires *task knowledge*, but it does not acquire *domain knowledge*. Consequently, if we change the goal of the task or if we change the agent's environment slightly, the learned policy may be completely worthless and the system may have to learn the modified task again from scratch.

Example 3.1. Figure 3.1 depicts two spatial situations with the only difference that they are differently oriented in the plane. One situation is rotated by 180° with regard to the other. System states are defined by coordinates $(x,y) \in \mathbb{R}^2$ in a global frame of reference and the robot's orientation $\theta \in [0,2\pi)$. The robot has already learned how to drive through the curve in the left environment. From the robot's point of view we would expect that after learning to solve the problem depicted on the left, the agent will also succeed in the turned scenario on the right. However, this is not the case. The strategy for the left situation is completely useless for the new task, because the semantics of each pair of (x,y,θ) and the learned action sequences do not lead to the destination (and not even to a sensible navigation behavior) when trying to succeed in the new problem.

One principle of reusing gained knowledge is *generalization*. It addresses the question of whether and how knowledge gained in a subset of $\mathcal{S}$ can be used in states outside this subset. In particular, generalization over states can mean that the system is able to infer that similar states have similar values, and following that, action selection will be the same in states that do not differ too much. In smooth continuous state spaces this is a property we certainly expect. A key question, of course, is how to specify "differ too much" in a given context.

So generalization allows for reusing experience from prior state samples and applying learned knowledge under slightly changed conditions. This can, on the one hand, refer to states that are very near with regard to a certain measure (such as coordinates very close to each other with regard to Euclidean distance in a continuous state space). In this case we speak of *local generalization*. On the other hand, it can also be the case for states that share a certain *structure* in the underlying domain. We define the structure of a domain as recurring properties that are significant for action selection for the task at hand. For example, the structure of a street map is given by roads, and indoor office environments are structured by walls. Such structural features influence action selection. In Example 3.1 the local structure of the walls induces a left turn by the agent independently of the global context. Many places in the environment may share this local structure, and that is where we want the agent to reuse its knowledge: If generalization covers structural commonalities, that is, if the system generalizes over states with the same structure, then we encounter a *non-local generalization*.

The question of knowledge reuse is coupled with the question of applicability of reinforcement learning to complex state spaces brought up in the previous section: If the learning system achieves an understanding of the structure of a state space, we can benefit from this understanding at various occasions and places spread over the whole state space. A system that is able to generalize does not need to learn everything anew from scratch when confronted with a new situation. On the contrary, gathered experience may be valid also in these situations, thus reducing learning efforts and reducing the part of the state space for which no knowledge has been acquired at all. Thus, reusable knowledge is key for real-world applicability of autonomously learned strategies.

The following sections give an overview of approaches to address the presented challenges in reinforcement learning.

3.2 Value Function Approximation

First, let us take a look at how the learned knowledge, that is, the value function, is stored. To represent V or Q a data structure is needed that associates s or (s,a) with a certain value. As long as the number of states is finite, the natural choice for that would be a multidimensional table. Accessing $V(s)$ or $Q(s,a)$ is then a simple table lookup or update. The table does not need to be able to hold a value for all theoretically possible system states, but only for the states visited within the learning experiment. This can be achieved with a hash table: values are stored compactly in a comparably small data array and a hash function returns the place where each $V(s)$ is stored for a state s.

However, as noted before, for large state spaces a tabular representation is infeasible. Realistic problems easily exceed the magnitude of millions of different system states, and, even worse, real-life problems usually do not have a discrete state space and have to be modeled as continuous-state MDPs. In these cases, a *value function approximation* method can be applied. The need for such compact representations of V when trying to solve complex problems is widely accepted (see Singh et al., 1995, for example).

The idea behind value function approximation is to find a compact representation that approximates the real value function as closely as possible. This is not only a drawback. On the contrary, tabular value function representations make inefficient use of experience especially in large state spaces (Kaelbling et al., 1996) because experience is stored sparsely over the table and information about neighboring or similar states is not regarded. In spite of that, value function approximation enables a key property of a learning system: *generalization*. So the mechanism of value function approximation covers both topics addressed in this section: handling large state spaces and reusing knowledge.

3.2.1 Value Function Approximation Methods

In principle, any function approximation method is applicable to reinforcement learning tasks. In the following, we will briefly look at some examples.

3.2.1.1 State Space Coarsening

A simple and straightforward method for storing values of a non-discrete set of states $\mathcal{S}$ is to pick a finite subset $\mathcal{T} \subset \mathcal{S}$ and to define a mapping function $\kappa : \mathcal{S} \rightarrow \mathcal{T}$ that assigns to each $s \in \mathcal{S}$ a representative $t \in \mathcal{T}$, for example, a function that maps a vector of real numbers component-by-component to the next smallest integer. For every state $s \in \mathcal{S}$ only the corresponding value of the representative $V(\kappa(s))$ is stored.

As the boundaries between the abstract states are arbitrary, this is not a reasonable method to approximate a value function. Generalization occurs over states that are mapped to the same entity, but only and exclusively within these states, while directly neighboring states might not be targets of generalization. For example, when rounding a value $s \in \mathbb{R}$ to the next smallest integer $\lfloor s \rfloor$, the value for $s = 1.9$ would map to 1 and would thus also affect $V(1.1)$, while $V(2)$ would not be affected at all, even though 1.9 is much closer to 2 than 1.1. The same holds if we use commercial rounding (mapping s to $\lfloor s + 0.5 \rfloor$) for $s = 1.4$ and $V(0.6)$ and $V(1.5)$. So state space coarsening is generally too rough to work as a successful value function approximation.

3.2.1.2 Coarse Coding

To address the mentioned shortcoming of the discretization described above, the concept of *coarse coding* makes use of a set $\mathcal{T}$ that consists of *overlapping* partitions of the state space. The state space is divided into non-disjoint sets, which are called

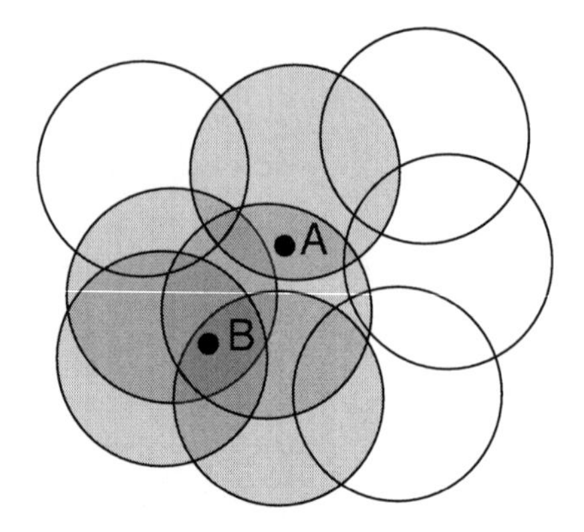

Fig. 3.2 Coarse coding (drawn after Sutton and Barto, 1998): Overlapping circular receptive fields partition the state space. State A is covered by two receptive fields, of which one also contains B. Thus, a slight generalization is achieved between A and B

receptive fields. For each receptive field a boolean function denotes whether a state is within it or not. The value $V(s)$ is then stored distributedly over the receptive fields. That is, we need to store values only for a finite set. Changing the value of a state s then changes all the values for all states that share one or more receptive fields with s. Thus, this update affects also states that have receptive fields in common with s. The more overlaps there are, the stronger the generalization. The effect of generalization—strong or weak, symmetrical or not—is influenced by the shape and size of the receptive fields. Figure 3.2 illustrates how coarse coding works.

3.2.1.3 Radial Basis Functions (RBFs)

A *radial basis function* (RBF) generalizes coarse coding to non-boolean features. Unlike a receptive field as defined above, an RBF feature has a continuous response in $[0,1]$ that is derived from a Gaussian function, for example. That means that a state is not within one of these features or not, but is within each of them to a certain degree. For a review on function approximation with RBFs see Powell (1987).

3.2.1.4 CMACs/Tile Coding

The *Cerebellar Model Articulator Controller* (CMAC) (Albus, 1981), which is also referred to as *tile coding* (Sutton, 1996), is a specific form of coarse coding. It has been frequently used for function approximation in reinforcement learning.

A CMAC consists of several different partitions of the state space, called *tilings*. A tiling is a tessellation of non-overlapping perceptive fields; in other words, every state s belongs to exactly one receptive field within each tiling. A receptive field in a tiling is called a *tile*. While usually in coarse coding the generalization effect is caused by the overlap of perceptive fields, the CMAC has disjoint tiles. But the tilings are organized in a way that each is arranged with a certain offset in each dimension to the others such that a state is mapped to different tiles in different tilings. The more the tilings, the finer the generalization. As tile coding plays an important role within the scope of this work, a detailed introduction to CMACs is provided in Sect. 5.3.2.

3.2.1.5 Artificial Neural Networks

CMACs and RBFs are *linear function approximations*, that is, the approximation is a linear function of the feature vector. These types of methods usually provide local generalization only, that is, structural properties of the state space are neglected.

For *nonlinear function approximators*, this restriction does not necessarily hold. A frequently used nonlinear method for function approximation is the feed-forward *artificial neural network* trained with backpropagation. Neural networks can give very good adaptation to the given input and provide a valuable generalization effect. One of the most famous applications to use feed-forward neural networks in combination with reinforcement learning is the widely recognized TD-Gammon program (Tesauro, 1992), which implements a very successful backgammon strategy based on TD(λ) learning. Reinforcement learning with backpropagation networks has also been applied in the robot soccer team of the University of Osnabrück (Gabel et al., 2006), which repeatedly earned the world champion title. Many more examples exist.

However, the drawback of neural networks is that learning in one part of the state space may lead to loss of knowledge in another part; this is referred to as the *inference problem* (Weaver et al., 1998). To cope with this problem is not easy, because of the "black box" character of neural networks. All in all, their usage is complicated and requires solid experience.

3.2.1.6 Other Methods

Function approximation is a critical issue, and not only in reinforcement learning research. Many elaborate methods have been and are still developed, for example, based on manifold representations (Glaubius and Smart, 2004), case-based reasoning (Gabel and Riedmiller, 2005) and evolutionary algorithms (Whiteson and Stone, 2006), to name a few.

A problem that arises when using linear function approximators is the choice of its basis, which has usually to be provided by the system designer. This is tackled by the seminal approach of Mahadevan (2005), which constructs the basis automatically after a global structural analysis of the state space based on Laplacian eigenfunctions that takes into account neighborhood relations of states. However, like most value function approximations, it assumes the value function to be smooth, that is, no sudden changes in the values of neighboring states have to be expected. But in practice we frequently encounter inhomogeneities in the value function, for example, near the walls of a corridor. The use of diffusion wavelets solves this problem (Mahadevan and Maggioni, 2006). However, this produces a high computational overhead (Petrik, 2007), and, as Sugiyama et al. (2007) point out, its application is not straightforward in practical use.

3.2.2 Function Approximation and Optimality

When function approximation is applied, TD methods in general lose the property of converging to an optimal value function, and often convergence cannot be guaranteed at all (Boyan and Moore, 1995). Not all types of function approximators show this unfortunate property: TD(λ) converges near an optimal solution when using linear function approximators (Tsitsiklis and Van Roy, 1997), but no convergence can be guaranteed for off-policy methods like Q-learning (Baird, 1995). Singh et al. (1995) could show that a certain kind of state aggregation (that is, clustering of different states) converges with a probability of 1 for Q-learning and TD(0). However, this kind of approximation is very limited.

Only very recently was the issue of convergence with function value approximation addressed again: Sutton et al. (2009b) introduced the first algorithm that converges reliably also for off-policy training with linear function approximators, the *gradient temporal difference* (GTD) algorithm. While the practical use of GTD is questionable due to its slow learning, *linear TD with gradient correction* (TDC) achieves the same properties as GTD with a learning speed comparable to ordinary TD-learning (Sutton et al., 2009a).

3.3 Temporal Abstraction

Besides value function approximation, a popular method to cope with the curse of dimensionality is the application of *temporal abstraction* to the problem specification and solution. In general, this means lowering the granularity of the discrete time scale.[1] Instead of action selection at every point in time, decisions are drawn less frequently, and temporally extended actions are invoked. This establishes a hierarchical control architecture, with extended actions on the higher level and simpler action primitives on the lower.

The usual formalism to describe this kind of problem is the so-called *semi-Markov decision process* (SMDP).

[1] Due to minor impact in reinforcement learning research, MDPs with continuous time scales are not addressed in this work.

3.3.1 Semi-Markov Decision Processes

A semi-Markov decision process (SMDP) (Howard, 1971) is a generalization of the MDP framework (see Sect. 2.3) that introduces the notion of time into the problem specification. While in an MDP actions are only considered as a sequence of decisions without any temporal specification, an SMDP takes the time needed to execute an action into account.

An SMDP defined on a discrete time scale is called a *discrete-time SMDP*. Here, decisions are only taken at positive integer multiples of the time scale of the underlying MDP, that is, actions are only selected every $n \in \mathbb{N}$ steps.

This corresponds to the interpretation that the system remains in each state $s \in \mathcal{S}$ for a certain time until the next action is invoked. This waiting time $\xi \in \mathbb{N}$ is another random parameter to be included in the transition probability T. In an SMDP, T is defined as $T : \mathcal{S} \times \mathcal{A} \times \mathcal{S} \times \mathbb{N} \to [0,1]$, and $T(s', a, s, \xi)$ describes the probability that in a state $s \in \mathcal{S}$ the selection of an action $a \in \mathcal{A}$ results in a state $s' \in \mathcal{S}$ after $\xi \in \mathbb{N}$ time steps.

Accordingly, the Bellman optimality equation for SMDPs is

$$V^*(s) = \max_{a \in \mathcal{A}} (R(s,a) + \sum_{s' \in \mathcal{S}, \xi \in \mathbb{N}} \gamma^{\xi} T(s,a,s',\xi) V^*(s')). \tag{3.1}$$

It enhances (2.3) by also iterating over the waiting time steps ξ and adapting the discount factor γ accordingly.

Most of the following approaches to temporal abstraction for reinforcement learning operate on the SMDP framework.

3.3.2 Options

To apply reinforcement learning to problems specified as SMDPs, Sutton et al. (1999) introduced the *options* model. An option subsumes a sequence of primitive actions that is derived by following a certain policy. Formally, it is a triple $\langle \mathcal{I}, \pi, \beta \rangle$ with an initiation set $\mathcal{I} \subset \mathcal{S}$, a policy π, and a termination condition $\beta : \mathcal{S} \to [0,1]$. An option can be executed if the system is in a state $s \in \mathcal{I}$, and then actions are selected according to π to reach a state s', where the option is terminated with a probability of β, or it continues. All primitive actions $a \in \mathcal{A}$ are also options with $\mathcal{I} = \mathcal{S}$ and $\beta(s) = 1$ for all $s \in \mathcal{S}$. Options can be defined hierarchically and thus provide a rich and complex architecture. The options model directly defines an SMDP such that all given theoretical and practical results of SMDP theory can be applied.

Options extend the initial MDP more than they simplify it, because options add to the possibilities of action selection, and thus the search space is even enlarged. However, the options model provides a good foundation for more sophisticated techniques. One example is *intra-option learning* (Sutton et al., 1998; Precup, 2000), which allows for learning within the execution of an option, and thus also enables updating other options that are partially identical, leading to a generalization effect.

Options can be provided a priori based on background knowledge. Their policies can be learned by standard RL methods as well. Therefore, the identification of subgoals within a task is an important, yet difficult issue. Subgoal detection is often based on heuristics, for example by identifying regions that are frequently visited during the learning process (McGovern and Barto, 2001).

3.3.3 MAXQ

A concept similar to the options model is the *MAXQ value function decomposition* (Dietterich, 1998, 2000a). One difference is that MAXQ creates a hierarchy of SMDPs that can be solved in parallel. The original MDP is divided into hierarchically organized subtasks, where—similarly to the options model—the actions (called children) of every subtask either are primitive actions or solve a subtask. Like an option, a subtask consists of three components: a set $\mathcal{S}_i \subseteq \mathcal{S}$ naming the states where a subtask can be executed, a policy π_i that selects an action from the subtask's children, and a pseudo-reward function. The pseudo-reward function allows for designing the system such in a way that subgoals can be defined, but without the need to specify the policy a priori. Additionally, a component is added to each state that maintains the complete history of calling subtasks in a stack data structure. Each subtask describes a discrete-time SMDP.

Dietterich presents an algorithm that recursively goes through the hierarchy of subtasks until primitive actions are executed, and the Q-values of the subtasks are updated after completing each call, taking into account the number of primitive actions executed to adapt the discount parameter γ. Remarkable is the fact that all subtasks are learned concurrently, without the need to wait for convergence of a subtask to start the learning of a higher level. The algorithm converges to a recursive optimal policy (Dietterich, 2000a).

Because MAXQ divides the overall value function into several value functions for subtasks, it is suitable for reuse of knowledge gained within the subtask. MAXQ enables powerful and efficient learning of hierarchical strategies. However, the key question of how to derive these hierarchies automatically is still open (Barto and Mahadevan, 2004).

3.3.4 Skills

An early approach to tackle the problem of applying reinforcement learning to more complex tasks was the introduction of *skills* (Thrun and Schwartz, 1995) as instances of temporal abstraction.

The authors have been among the first to note that for useful application of RL the structure of the state space has to be taken into account. A skill k, as they define it, is a sequence of actions that is locally defined over a subset $\mathcal{S}_k \subset \mathcal{S}$ and collapses this sequence of actions into a single operation. Skills are not specific to a single task, but apply to several different ones where they are to replace the local action policies. This abstraction, of course, comes at the price of a loss in performance with regard to the optimality of the learned policy. The proposed SKILLS algorithm to derive skills considers commonalities in policies over a single state space.

The skills approach shows promising performance improvements in grid world domains. However, the authors note that the derivation of useful skills requires time up to an order of magnitude larger than that of finding a near-optimal policy.

The SKILLS algorithm works as an a posteriori analysis of a learned strategy. Thus, skills can be used for generalization across different tasks, but cannot generalize over structurally similar state sets within a given task.

3.3.5 Further Approaches and Limitations

All the presented approaches to temporal abstraction are similar: They collapse sequences of actions into one singular macro action. Another approach following this concept is the *hierarchy of abstract machines* (HAM) (Parr and Russell, 1998), which is of minor importance with regard to larger-scale applications. In general, large-scale environments are still a challenge for the application of the named hierarchical temporal abstraction techniques (Barto and Mahadevan, 2004). One basic drawback of these approaches is the need for complete observability of the state space. Dietterich (2000b) illustrates this by the intriguing everyday example that "when deciding how to walk from the bedroom to the kitchen, we do not need to think about the location of our car" and concludes that temporal abstraction methods, provided as they are, are unable to scale.

Another challenge is the automatic creation and adaptation of hierarchies, as stated above for MAXQ, which also holds for the options model and for similar approaches such as *hierarchical Q-learning* (HQL) (Kirchner, 1998). Furthermore, the problem of dynamic abstraction is still unsolved. A good summary on open issues of hierarchical reinforcement learning is given in Barto and Mahadevan (2004).

3.4 Spatial Abstraction

In contrast to abstracting the action level of a problem, one can also take the states of the system into account. Applying abstraction to a state space $\mathcal{S}$ means collapsing sets of different states $s \in \mathcal{S}$ into indistinguishable *abstract* states $\omega \in \mathcal{O}$. Thus, we create an observation space where several original states share the same observation (see Sect. 2.3.3 for partial observability). In this case, we speak of an *abstract state space*. The size of the state space is effectively reduced as several states in the original state space can no longer be distinguished. This contrasts with the approaches based on value function approximation (Sect. 3.2), where the original state space is maintained such that every state can be distinguished and accessed, and only the storage of the value function is subject to abstraction.

In the very simple case of value coarsening (Sect. 3.2.1.1), state space abstraction and function approximation are equivalent: It does not make a difference whether two states s and s' share the same value because the value is $V(\omega)$ with $\omega = \psi(s) = \psi(s')$ or because $V(s) = V(s') = V(\kappa(s)) = V(\kappa(s'))$. For non-trivial function approximation methods, of course, value function approximations and spatial state space abstraction denote different concepts.

The main question in spatial abstraction is which states are aggregated to an abstract state. The key to sensible abstraction is again the structure of the underlying state space. In the following section we take a look at how structure is utilized in common state space abstraction approaches.

3.4.1 Adaptive State Space Partitions

One way to find a smart representation of the state space according to its structure is to start from a coarse partition which is refined adaptively according to the underlying structural properties. Several such approaches have been presented so far.

The parti-game algorithm (Moore and Atkeson, 1995) was introduced as one of the first approaches to automatically create an abstract state space. It has been presented for goal-directed navigation problems in 2-D, but the authors show that it also can be applied to higher-dimensional problems. The state space is divided into cells. While the agent is operating, these cells are incrementally divided to achieve a higher resolution. The necessity of a cell partition is determined by a minimax criterion that detects whether the current spatial resolution of the grid in this area prohibits a successful state transition between obstacles. A local controller is needed as part of the problem definition to achieve the transitions from cell to cell. Thus, the grid's resolution is extremely fine near obstacles and rough in open spaces of the environment.

The parti-game algorithm has become very famous and has been adapted and improved multiple times (see, for example, Al-Ansari and Williams, 1999; Likhachev and Koenig, 2003). It also motivated real robot navigation approaches such as the variable-resolution cognitive maps (Arleo et al., 1999). However, its practical importance is limited because the applicability of the algorithm is restricted to deterministic problems which we almost never encounter in realistic settings. Also, it suffers from the restriction that the size of the environment has to be known in advance.

Another adaptive state space partitioning is described by Dean and Givan (1997), who present an approach for minimizing the model of an MDP. They propose partitioning the world homogeneously into groups of states (called aggregates) within which all states share the same transition probabilities. But in large continuous domains—and it is this kind of domain that is relevant for most practical applications—this still results in overly complex representations. Furthermore, knowledge about the model of the system is needed a priori, so this method cannot be applied to model-free learning algorithms such as Q-learning. This is a serious drawback, because a model of the system if often not available, and the fact that a model is not needed to achieve an optimal result is one of the main reasons for employing autonomous learning on such problems.

The same dependency on a known model holds for the so-called variable resolution discretization (Munos and Moore, 1999). Starting from an initial coarse grid approximation of the state space, this partition is incrementally refined in areas where it is necessary, thus approximating the system dynamics up to a certain level of exactness. As a data structure to represent these refinements, a *kd-tree* is used.

A model-free approach to partitioning the state space has been introduced by Reynolds (2000). Also using kd-trees, the cells of a coarse initial partition are split up. To determine which cells to split in which way, so-called *decision boundaries* are considered: If the two subcells that emerge after partitioning show different optimal actions according to their Q-values, then the split will be performed. So the algorithm tries to create regions with uniform optimal action selection. That means that theoretically the optimal policy π^* must already be known to start the partitioning process; this is not the case, as π^* is the result of the learning process. However, Reynolds claims that near-optimal policies may suffice to determine decision boundaries, and that those policies can often be derived also on a coarse state space partition. If this assumption does not hold, the presented method does not work.

kd-Q-learning (Vollbrecht, 2000) is another model-free approach to discretizing the state space using kd-trees. Here, a kd-tree provides several different resolutions of the state space at the same time which are simultaneously maintained over the learning process. This method has proved to accelerate reinforcement learning, especially in the early learning phase. However, it requires that it be possible to hold the complete representation of the finest abstraction level in memory—that is, a table representation of the value function must be possible. Furthermore, it can not inhibit state explosion, so the algorithm is hardly applicable to more complex problems.

The continuous U-Tree approach (Uther and Veloso, 1998) extends the U-Tree algorithm (McCallum, 1995) to continuous state spaces. Here, current and past observations in the world are cumulated in abstract states. Starting from a single state, they are split according to statistical tests. However, splits can only be performed along the axis of the input space, and more flexible splitting abilities are a matter of extended efforts. The authors claim continuous U-Tree to be suitable for moderately high-dimensional state spaces.

The BASM algorithm (Timmer and Riedmiller, 2006) can be applied to discretizations of the continuous state space of an MDP to gain a near-optimal solution. It enriches abstract states by a history of prior states and transitions to tackle the state aliasing problem. Starting from a few abstract states, they are incrementally refined on the basis of statistical tests that detect significant differences in the distribution of Q-values. This implies that the refinement can only take place if the sampled Q-values are more or less meaningful, that is, the resulting policy at this point must already show certain success. Thus, as in Reynolds (2000), deriving a near-optimal policy must be possible on a coarse partition.

3.4.2 Knowledge Reuse Based on Domain Knowledge

Instead of relying on a model of the system or a suboptimal policy on a coarse state space and adaptively reshaping the state space representation, we can also use domain knowledge directly to specify a suitable spatial representation or relations between states. In this section, we investigate some of these approaches. Some of them also rely on value function approximation or temporal abstraction. Because the application domain considered in this book is robot navigation, we primarily look at methods operating in this domain.

Topological neighborhood relations within the state space are used by Braga and Araújo (2003) to improve the learning performance of a robot navigation task. The reinforcement signal is propagated also through neighboring states in the environment, allowing knowledge to be shared over related states. This approach requires an a priori existence of a topological map of the environment the robot is operating in.

Glaubius et al. (2005) concentrate on the internal value function representation to reuse experience across similar parts of the state space. They also need a priori knowledge in form of predefined locally similar equivalence classes to distinguish between similar regions in the world, represented in a value function approximation based on manifolds (Glaubius and Smart, 2004).

Lane and Wilson (2005) describe relational policies for spatial environments and conditions under which they can be "approximately relocated" within the learning task. Therefore, they rely on distance and orientation between states. The authors demonstrate significant learning improvements in goal-directed robot navigation tasks. However, their approach is only suitable for homogeneous environments, such as open spaces of outdoor fields, and runs into problems when non-homogeneities such as walls and obstacles appear. To avoid that shortcoming, the authors suggest taking into account the relative position of walls with respect to the agent, but to my knowledge, they have not yet implemented this idea.

Partial views and *partial rules* operating on them are introduced in Porta and Celaya (2005) for using both spatial and temporal abstraction in a robot navigation task if it matches the condition of *categorizability*—that is, if "a reduced fraction of the available inputs and actuators have to be considered at a time." Partial rules have to be given initially and are refined when a larger deviation in the value prediction is encountered. Also here, a considerable amount of a priori knowledge is needed to derive the partial rules.

3.4.3 Combining Spatial and Temporal Abstraction

Because of the suboptimal scalability of temporal abstraction approaches as pointed out in Sect. 3.3.5, spatial abstraction is also increasingly incorporated into these methods.

Dietterich (2000b) presents an extension of the MAXQ framework to spatial abstraction and is even able to prove convergence to optimality. He concludes that spatial abstraction is important for successful application of MAXQ.

Also, spatial abstraction techniques can be enriched by a temporal abstraction component: The *TTree* algorithm (Uther and Veloso, 2003) extends the continuous U-tree algorithm (see Sect. 3.4.1) by policies (that is, solutions of subtasks) as temporally abstract actions that repeat the same primitive action until the next abstract state is reached or performs random moves. This results in an abstract SMDP. The policy solving the abstract SMDP can then be mapped back to the base SMDP.

3.4.4 Further Task-Specific Abstractions

An abstract spatial representation tailor-made for the task at hand is presented in Busquets et al. (2002) to extend an approach to coordinate the actions of robots in a multi-agent setting (Sierra et al., 2001). The authors use a component that autonomously learns rules using prioritized sweeping (Moore and Atkeson, 1993), a model-based reinforcement learning method. This approach uses an egocentric qualitative abstraction and fuzzy numbers to represent the whereabouts of detected landmarks around a robot. The navigation relies on rather complex actions, and obstacle avoidance is handled by a separate component. The scenarios considered in this book provide much simpler motion models with basic action primitives and without high-level modules such as collision avoidance.

Also based on the knowledge of a model, Theocharous et al. (2005) explicitly study robot navigation as an instance of the POMDP class. They present the H-POMDP algorithm that uses both temporal and spatial abstraction within a complex hierarchical framework, using EM algorithms like the Baum-Welch algorithm (Baum and Sell, 1968). Their approach shows good performance compared to other POMDP algorithms. However, it requires a good initial model of the POMDP in the beginning.

3.5 Transfer Learning

Generalization is about how gained knowledge can be reused within the current learning task. The question of how a learned policy can be beneficial within a new task, within a new environment, and/or with a new goal, has recently been subsumed

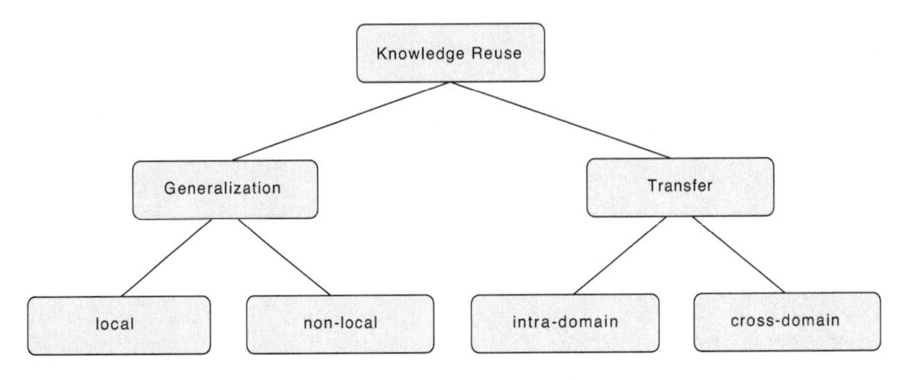

Fig. 3.3: Overview: different facets of knowledge reuse in reinforcement learning

under the (slightly misleading) name *transfer learning*. It can be seen as a form of "cross-task generalization."

The question of whether "learning the n-th thing [is] any easier than learning the first" has been asked by Thrun (1996), who investigated the problem of knowledge reuse in a context of life-long learning. For a human, we expect transfer learning to be a natural thing: The knowledge of how to drive a passenger car, for instance, can immediately be applied to a certain extent if the driver is asked to drive a truck, because several aspects of these tasks are identical, such as the position of the steering wheel. Humans are very good at such transfer efforts. In machine learning, and in particular in reinforcement learning, this does not happen naturally.

Transfer learning can be divided into two differently challenging categories: approaches that solve new tasks within a given domain and approaches that work in different domains. The former is called *intra-domain transfer* and the latter *cross-domain transfer* (Taylor and Stone, 2007). We speak of the same domain if state space, action space, and transfer probabilities are identical, but the task to solve is a different one. That is, intra-domain transfer tackles problems that differ in the reward function.

Figure 3.3 shows an overview of the different types of knowledge reuse in autonomous learning problems: local and generalization as well as intra- and cross-task transfer.

3.5.1 The DARPA Transfer Learning Program

The United States Defense Advanced Research Projects Agency (DARPA) initiated a transfer learning program in 2005 boosting transfer learning efforts by scientists worldwide. The DARPA defines transfer learning methods in its Broad Agency Announcement (BAA) 05-29 as "techniques that enable computers to apply knowledge learned for a particular, original set of tasks to achieve superior performance on new, previously unseen tasks" (Defense Advanced Research Projects Agency, 2005).

BAA 05-29 also defines an 11-level hierarchy of transfer learning, starting from level 0 ("memorization," no transfer, improvement in identical tasks) to level 10 ("differing," using common knowledge from training instances that show minimal overlap with the new problem). Levels 0 to 9 are only regarded as intermediate steps to level 10, which will solve the transfer learning problem. The hierarchy gives a good overview on the complexity of the problem and how much effort is needed to succeed in transfer learning. However, the definitions of the separate levels remain vague, which may be the reason that, in spite of the major impact of the DARPA project to transfer learning research, this hierarchy is hardly ever referenced in scientific publications. Thus, we do not go into detail here.

3.5.2 Intra-domain Transfer Methods

The temporal abstraction methods described in Sect. 3.3 all have in common that they collapse sequences of primitive actions into single extended actions, and the approaches are organized in a hierarchical manner. On the lower levels of the hierarchy, certain subgoals are solved. These subgoals usually do not only refer to the solution of one overall problem, but may be useful for a larger set of tasks: For example, the subtask "leave a room and enter the corridor" can be useful for handling a navigation task in any indoor environment.

Thus, all temporally abstracted solutions may be used for intra-domain transfer. As the state space stays the same, the extended actions can be used also in new tasks within this domain if known subgoals have to be reached also in the new task. This possibility is independent of the temporal abstraction paradigm, be it options (Sect. 3.3.2), HAM (Sect. 3.3.5), MAXQ (Sect. 3.3.3), or any other approach based on temporal abstraction and subgoals. The skills framework (Sect. 3.3.4) was explicitly designed as an instance of intra-domain transfer.

The same holds for *policy reuse* (Fernández and Veloso, 2006), which makes use of the experience of past learned policies to solve different tasks within the same domain more efficiently. The method consists of a biased exploration method based on probabilistics and a similarity function to judge usefulness of past policies. Additionally, policies that are different enough are stored in a policy library for later use.

3.5.3 Cross-domain Transfer Methods

Source and target task may be so similar in structure that they can be considered homomorphous. Ravindran (2004) introduced MDP homomorphisms as a concept of abstraction to transfer one MDP to another MDP. An MDP homomorphism is a mapping $h : \mathcal{S} \times \mathcal{A} \rightarrow \mathcal{S}' \times \mathcal{A}'$ that transforms the base MDP $\langle \mathcal{S}, \mathcal{A}, T, R \rangle$ into a new MDP $\langle \mathcal{S}', \mathcal{A}', T', R' \rangle$ such that certain conditions between R' and R as well as between T' and T hold. If we encounter homomorphous MDPs, then policies can easily be transferred between the MDPs based on the homomorphism. However, problems with such a high similarity are rare, and not the main focus of transfer learning research. Also, discovering MDP homomorphisms is NP-hard (Ravindran and Barto, 2003).

But the MDP homomorphism framework can also be used in transfer learning approaches, for example, Wolfe and Barto (2006b), where they are used to identify objects of the same type, for which options (see Sect. 3.3.2) are constructed that can be reused for all of these objects. This, of course, is an intra-domain transfer method. It is enhanced by combining it with the U-tree algorithm (see Sect. 3.4.1) to find reusable MDP homomorphisms (Wolfe and Barto, 2006a). Another approach to use MDP homomorphisms and intra-option learning for cross-domain transfer is given by Soni and Singh (2006).

The AI^2 algorithm (Torrey et al., 2006) works on target tasks that operate in the same environment, but with different rewards, actions, and state variables. It constructs so-called *transfer advice* by analyzing policies in the source task with inductive logic programming and describing skills in first-order logic. When learning the target task, this advice is considered. The mapping procedure between the two tasks has to be given by the user a priori.

Taylor and Stone (2007) present the *rule transfer* algorithm that summarizes a learned policy into a set of rules, the so-called decision list. The decision list is an abstract representation of the source task policy that is then modified to match the target task. This modification uses predefined translation functions, that is, the connection between source and target task is given a priori and based on external knowledge. This leveraging of rules results in faster learning in a target task in a different domain. As in AI^2, the relationship between source and target task has to be specified by an expert.

The latter two approaches are enhancements of prior approaches (Torrey et al., 2005; Taylor and Stone, 2005) that both transferred a Q-function directly instead of using abstract rules or first-order logic. Those earlier approaches were limited to cases where source and target task are very similar, and the authors claim to achieve better results with the abstract transfer algorithms. Again, the relationship between the tasks has to be provided by an expert. A method to automatically derive these methods is given by Liu and Stone (2006): Based on a model of source and target task in the form of a qualitative dynamic Bayesian network (QDBN), the mapping can be derived automatically. However, a model of both source and target task has to be present a priori to derive these QDBNs.

Konidaris and Barto (2007) show how options can be transferred from one domain to another based on the transfer learning framework of *problem space* and *agent space* (Konidaris, 2006). The approach would also work with MAXQ or HAM instead of options. This transfer learning framework, which has been developed concurrently to the present work, presents ideas and conclusions closely related to the research presented in this book. Thus, it is described in greater detail, along with the presentation of a new abstraction-based approach to transfer learning, in Chap. 5.

3.6 Summary and Discussion

For successfully applying reinforcement learning to complex scenarios, abstraction is a key issue. On the one hand, abstraction allows for handling an overly complex state space; on the other hand, it can allow the system to reuse previously gathered knowledge.

Value function approximation allows for storing the approximated value function in a compact representation. In most of the cases, this only leads to local generalization. However, for coping with larger domains global generalization is desirable, and the key to global generalization is the notion of structure within the state space. However, function approximation methods utilizing structural properties are difficult to apply.

When it comes to knowledge reuse, mostly the paradigm of temporal abstraction is applied to derive options or other sequences of primitive actions that can be used at different places within the state space or in new tasks. However, the applicability to larger state spaces is limited, so the abstraction of the state space becomes important. A sensible spatial abstraction can be the basis for applying advanced algorithms. This book wants to put the focus on the question of how such a sensible abstraction has to look and how it can be derived. When abstaining from temporal abstraction, the flexibility in the agent's motion dynamics can be preserved.

A sensible abstraction is related to the structure of the state space. This has often been implemented by state space partitionings that try to derive a structure while learning or by integration of domain knowledge into the state representation. However, with exceptions, such as the work of Glaubius et al. (2005), these approaches only achieve local generalization.

Neither temporal nor spatial abstraction has been applied successfully without incorporating domain knowledge. Also, all existing transfer learning approaches require the a priori definition of relationships between source and target task. The claim this book makes is that in previous approaches structural knowledge about the task is incorporated too late in data processing from sensor to value function. Especially, the relationship between state space representation and non-local generalization has been insufficiently illuminated. The question that arises here is how to use structural knowledge to make sure that the agent is confronted with an observation of the surrounding world that explicates the structural properties of the state space and thus enables it to act reasonably and learn efficiently. Also, a good representation should allow for immediately performing cross-domain transfer, that is, for using existent knowledge in new situations without the need for a complex transfer process. In other words, the problems raised within this chapter can be tackled by the use of an appropriate abstract state space representation.

So, in the next chapters, we will derive principles and algorithms on how to construct a state space abstraction that allows for successful learning of problems with large continuous state spaces and also enables both non-local generalization and easy and straightforward cross-domain transfer of learned policies based on the structure of the state space.

Chapter 4
Qualitative State Space Abstraction

In this chapter, we take a deeper look into the nature of abstraction. In particular, this chapter argues for the use of state space abstraction (Sect. 4.1) and presents a formal framework of abstraction and its different facets in Sect. 4.2. Based on this framework, a view on the interdependence of abstraction and representation follows in Sect. 4.3. Section 4.4 examines abstraction in agent control processes before applying them to reinforcement learning (Sect. 4.5). The use of qualitative representations for abstraction is argued for in Sect. 4.6 before the chapter ends with a summary.

4.1 Abstraction of the State Space

Abstraction is one of the key capabilities of human cognition. It enables us to conceptualize the surrounding world, build categories, and derive reactions from these categories to cope with different situations. Complex and overly detailed circumstances can be reduced to much simpler concepts, and not until then does it become feasible to deliberate about conclusions to draw and actions to take.

As described in Chap. 3, abstraction is also a key issue in machine learning for successfully and efficiently learning complex navigation tasks (Thrun and Schwartz, 1995). It can be applied to both action space $\mathcal{A}$ (see Sect. 3.3) and state space $\mathcal{S}$ (see Sect. 3.4). This book argues for applying abstraction to the state space to address the challenges of large continuous state spaces and knowledge transfer in reinforcement learning. While temporal abstraction does not scale well (Dietterich, 2000b) with the complexity of the problems, we concentrate on spatial abstraction for the following additional reasons:

- Sensor information is rich and provides lots of unnecessary information, which easily triggers the curse of dimensionality (see Sect. 3.1.1). Abstraction reduces the amount of data. Human cognition has developed mechanisms to focus on relevant information instead of treating all pieces of information equally, and this

L. Frommberger, *Qualitative Spatial Abstraction in Reinforcement Learning*, Cognitive Technologies, DOI 10.1007/978-3-642-16590-0_4,

selection of certain *features* is then used to reason about actions to take. Thus, the aim is to provide the learning agent with a vector of inputs that is relevant to derive a successful strategy.
- Sensible abstraction can help stress relevant details over irrelevant ones. Even if it is important for action selection, the level of importance of a detected feature can vary. For example, a wrong turn of a navigating agent at a decision point may lead to longer travel, but a wrong turn near an abyss can destroy the agent physically, so the abyss has to be considered critical. Such considerations can be part of the abstraction process.
- Spatial abstraction facilitates generalization. When different states are indistinguishable, the selection of action has to be identical; this provides generalization over the underlying states. If the indistinguishability also holds for similar places across the state space or even different domains, non-local generalization and cross-domain transfer become possible.
- The motion dynamics of the agent remain unchanged. When the action space is not subject to abstraction, the agent is capable of applying the full range of actions at any point in time, allowing for maximal flexibility. An appropriate spatial abstraction must provide enough distinguishable observations at places where this flexibility is needed.
- Easy integration into existing approaches can be provided, because it is just the representation of the states that changes. Different techniques to improve the learning process, such as modified or better learning algorithms, function approximation, exploration techniques, and so on, can still be used under abstraction of the state space.

Before answering the question of how to achieve these benefits, we start by taking a formal look at abstraction.

4.2 A Formal Framework of Abstraction

In their thorough study on abstraction, which they refer to as *schematization*, Klippel et al. (2005) state that “there is no consistent approach to model schematization.” With the exception of the formal definitions of abstraction principles on graphs given by Stell and Worboys (1999), most scientific work on abstraction remains informal or based on examples. However, it is the author’s belief that only a consistent formalization of abstraction allows for a thorough investigation of its properties and effects. Thus, in the following sections, we take a computer scientist’s view on abstraction to provide a formal framework of abstraction. We distinguish between three different facets of abstraction, namely *aspectualization*, *coarsening*, and *conceptual classification*. A consideration of the relationships of the facets and their

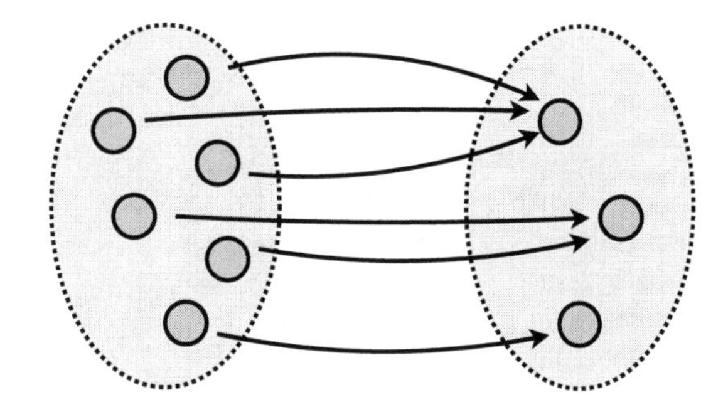

Fig. 4.1 Abstraction is a non-injective mapping: several different states collapse into identical entities in the target space

implication on agent control processes then allows for deriving principles of state space abstraction for reinforcement learning.[1]

4.2.1 Definition of Abstraction

The term *abstraction* is etymologically derived from the Latin words "abs" and "trahere," so the literal meaning is "drawing away." However, when talking about abstraction in the context of information processing and cognitive science, abstraction is more than just taking away something arbitrarily, because it is not merely intended to reduce the amount of data. Such a reduction could also be achieved by a defective data cable, but this is not an intentional process. Rather, abstraction puts the focus on the *relevant* information. Additionally, the result is supposed to generalize and be useful for a specific task at hand. We define abstraction as follows:

Definition 4.1. Abstraction is the process or the result of reducing the information of a given observation in order to achieve a classification that omits information irrelevant for a particular purpose.

We first concentrate on information reduction. Let us say that all potential values of a knowledge representation are elements of a set $\mathcal{S}$ which can be regarded as a Cartesian product of features from different domains: $\mathcal{S} = \mathcal{D}_1 \times \mathcal{D}_2 \times \ldots \times \mathcal{D}_n$. We call $s = (s_1, \ldots, s_n) \in \mathcal{S}$ a *feature vector*, and every s_i is a *feature value*.

Abstraction is a non-injective function $\kappa : \mathcal{S} \to \mathcal{T}$ mapping the source space $\mathcal{S}$ to a target set $\mathcal{T}$, that is, at least two states $s, s' \in \mathcal{S}$ are mapped to one $t \in \mathcal{T}$. In other words, abstraction is an n:1-mapping from $\mathcal{S}$ to $\mathcal{T}$ (see Fig. 4.1). An injective function κ would be no abstraction, as no information reduction can be achieved. Non-injectiveness is important as otherwise no reduction is possible. In the case of $\mathcal{S}$ being finite it holds that $|\mathcal{S}| > |\text{Image}(\kappa)|$. Without loss of generality we assume in the following, simply to facilitate readability, that all domains $\mathcal{D}$ are of the same kind, that is, $\mathcal{S} = D^n$. Thus, each $s \in \mathcal{S}$ can be represented as a vector

[1] Sections 4.2 to 4.4 present a significantly revised and extended version of the formal framework of abstraction published in Frommberger and Wolter (2008).

$s = (s_1, s_2, \dots, s_n)$, $s_i \in \mathcal{D}$. However, all the following considerations can also be formulated for a general state space $\mathcal{S}$. However, we assume that $\mathcal{S}$ is of finite dimension.

We also assume in this work that the representation of s is not redundant, that is, there exists no function $f : \mathcal{D}^k \rightarrow \mathcal{D}^l$ that maps parts of s to other parts of s. For example, $(s_1, s_2, s_3) = (x, y, x+y)$, $x \in \mathcal{D}$ is a redundant representation because s_3 is expressed as a function of s_1 and s_2. It might be impossible to ensure that a representation is free of internal redundancies. But for the following theoretical considerations, it is assumed that the representation is not deliberately designed to be redundant.

Abstraction can have several forms. We can distinguish between three different facets of abstraction. For example, it is possible to regard a subset of the available information only, or the level of detail of every piece of information can be reduced, or the available information can be used to construct new abstract entities.

Various terms have been coined for abstraction principles, distributed over several scientific fields like cognitive science, artificial intelligence, architecture, linguistics, geography, and many more. Among other terms we find *granularity* (Zadeh, 1979; Hobbs, 1985; Bittner and Smith, 2001), *generalization* (Mackaness and Chaudhry, 2008), *schematization* (Klippel et al., 2005; Herskovits, 1998), *idealization* (Herskovits, 1998), *selection* (Herskovits, 1998; Stell and Worboys, 1999), *amalgamation* (Stell and Worboys, 1999), or *aspectualization* (Berendt et al., 1998; Bertel et al., 2004, 2007). Unfortunately, some of these terms define overlapping concepts, different terms sometimes have the same meaning, or a single term is used for different concepts. Also, these terms are often not distinguished between in an exact manner or are only defined by examples.

This work studies abstraction as part of knowledge representation. The primary concern is representation of spatial knowledge in a way that maintains a perspective as general as possible, allowing adaptation to other domains.

The following sections distinguish between the three different facets of abstraction: *aspectualization*, *coarsening*, and *conceptual classification*.

4.2.2 Aspectualization

An *aspect* is a semantic concept. It is a piece of information that represents a certain property. For example, if we record the trajectory of a moving robot, we have a spatio-temporal data set denoting at what time the robot visited which place. Time and place are two different aspects of this data set. Aspectualization singles out such aspects. It is defined as follows:

Definition 4.2. Aspectualization is the process or result of explicating certain aspects of an observation purely by eliminating the others. Formally, it is defined as a function $\kappa : \mathcal{D}^n \rightarrow \mathcal{D}^m (n, m \in \mathbb{N}, n > m)$:

$$\kappa(s_1, s_2, \dots, s_n) = (s_{i_1}, s_{i_2}, \dots, s_{i_m}),\ i_k \in [1, n],\ i_k < i_{k+1}\ \forall k. \tag{4.1}$$

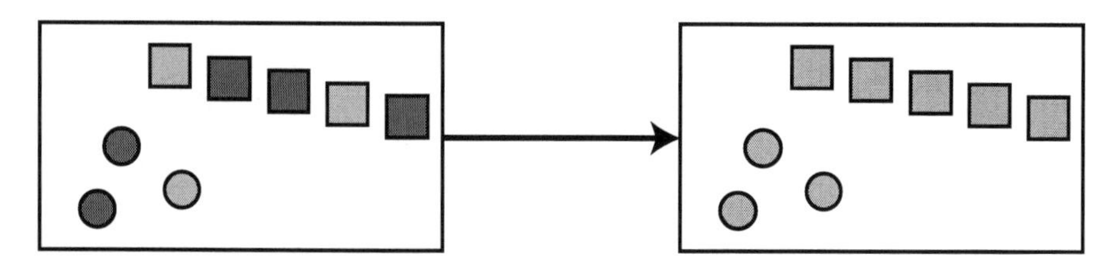

Fig. 4.2: Aspectualization abstracts from certain aspects. In this case, information on color is omitted

Aspectualization projects $\mathcal{D}^n$ to $\mathcal{D}^m$: The dimension of the state space is reduced. This elimination of feature dimensions contrasts with other possible methods of highlighting aspects, for example, by moving them to the beginning of the feature vector or by duplicating the corresponding features.

Figure 4.2 depicts aspectualization, and the following example illustrates it:

Example 4.1. An oriented line segment s in the plane is represented as a point $(x,y) \in \mathbb{R}^2$, a direction $\theta \in [0,2\pi]$, and a length $l \in \mathbb{R}$: $s = (x,y,\theta,l)$ (see Fig. 4.3). The reduction of this line segment to an oriented point is an aspectualization with $\kappa_a(x,y,\theta,l) = (x,y,\theta)$.

Aspects may span several features s_i. However, to be able to single out an aspect from a feature vector by aspectualization, it must be guaranteed that no feature refers to more than one aspect. We call this property *aspectualizability*.

Definition 4.3. If an aspect is exclusively represented by one or more components of a feature vector $s \in \mathcal{S}$ (that is, a subset of $\{s_1,\dots,s_n\}$ represents this aspect completely and does not refer to more than this aspect), then we call $\mathcal{S}$ aspectualizable with regard to this aspect.

From this definition, we can directly derive that if a state space $\mathcal{S}$ is aspectualizable with regard to an aspect, it is also aspectualizable with regard to everything but this aspect.

It is possible to find an aspectualizable and a non-aspectualizable representation that have the same semantics, as the following example shows:

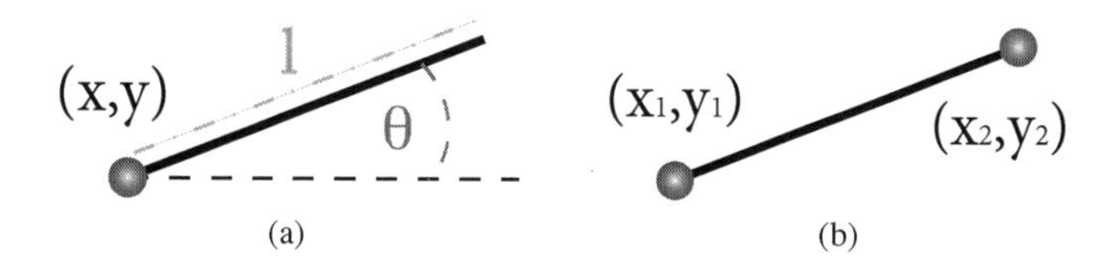

Fig. 4.3: Line segments in $\mathbb{R}^2$ can be defined by a point, an angle, and a distance (a) or by two points (b). Variant (a) is aspectualizable with regard to length; variant (b) is not

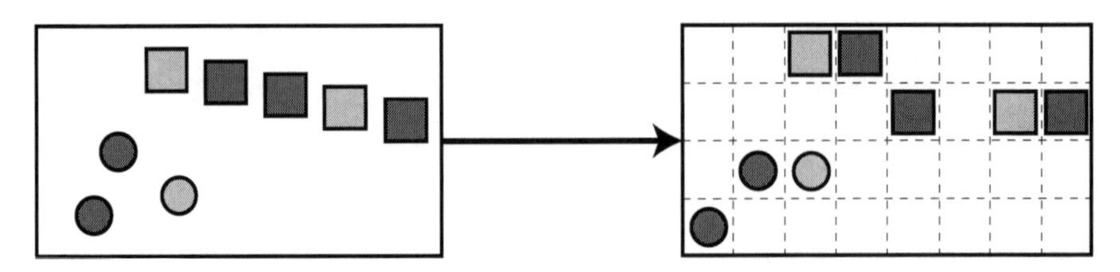

Fig. 4.4: Coarsening reduces the resolution of feature values. In this example, coarsening abstracts from the coordinates of the spatial entities by arranging them into a grid map

Example 4.2. The oriented line segment representation in Example 4.1 (Fig. 4.3a) can be bijectively mapped from point, angle, and length to two points (x_1, y_1, x_2, y_2) (Fig. 4.3b). Then aspectualization as defined in Def. 4.2 cannot single out the length of the line segment, because length is not represented explicitly. It is implicitly given and can be calculated from $((x_2, y_2) - (x_1, y_1))$, but all the coordinates are also needed to define the direction of the oriented line. Thus, $\mathcal{S}$ is not aspectualizable regarding length.

4.2.3 Coarsening

Aspectualization reduces the number of features, but retains the co-domains of the remaining ones. When the set of values a feature dimension can take is reduced, we speak of a coarsening:

Definition 4.4. Coarsening is the process or result of reducing the details of information of an observation by lowering the granularity of the input space. Formally, it is defined as a function $\kappa : \mathcal{D}^n \to \mathcal{D}^n (n \in \mathbb{N})$,

$$\kappa(s) = (\kappa_1(s_1), \kappa_2(s_2), \ldots, \kappa_n(s_n)) \tag{4.2}$$

with $\kappa_i : \mathcal{D} \to \mathcal{D}$ and at least one κ_i being not injective.

The required existence of a non-injective κ_i is necessary to have an abstraction according to Def. 4.1. For a finite set it $\mathcal{S}$ holds that $|\mathcal{S}| = |\text{Image}(\kappa)|$. Figure 4.4 depicts coarsening.

Definition 4.4 does not require any homogeneity in the mapping. Many coarsenings, however, show such a property. Thus, we also define a *homogeneous coarsening* for the case where we have an ordering $\leq$ defined on $\mathcal{D}$:

Definition 4.5. We have a coarsening κ and $x, y, z \in \mathcal{S}$. κ is called a *homogeneous coarsening* if for all x, y with $\kappa(x) = \kappa(y)$ and each component $i \in \{1, \ldots, n\}$ it holds that $x_i \leq z_i \leq y_i \Rightarrow \kappa_i(x_i) = \kappa_i(z_i)$. If for all x, y with $\kappa(x) = \kappa(y)$ it holds that $x_i \leq z_i \leq y_i \Leftrightarrow \kappa_i(x_i) = \kappa_i(z_i)$, then κ is called *totally homogeneous*.

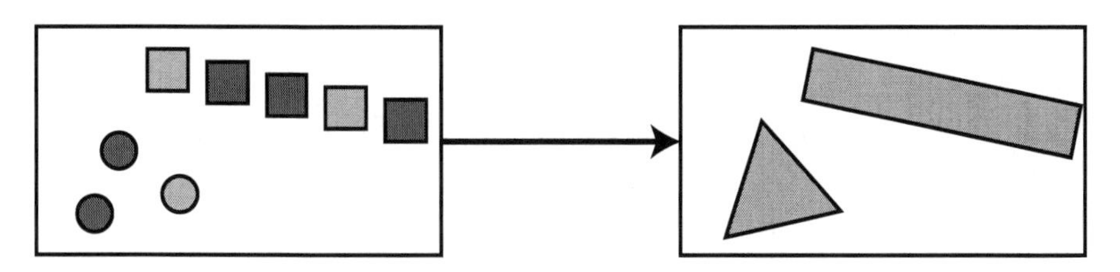

Fig. 4.5: Conceptual classification groups features to form completely new abstract entities

Example 4.3. An important representation in the area of robot navigation is the occupancy grid (Moravec and Elfes, 1985), a partition of 2-D or 3-D space into a set of discrete grid cells. A function $\kappa : \mathbb{R}^2 \to \mathbb{R}^2, \kappa(x,y) = (\lfloor x \rfloor, \lfloor y \rfloor)$, is a coarsening that maps any coordinate to a grid cell of an occupancy grid. This establishes a state space coarsening as described in Sect. 3.2.1.1. Also, κ is totally homogeneous.

4.2.4 Conceptual Classification

Conceptual classification is the most general of the three proposed abstraction facets. It can use all components of the input to build new entities:

Definition 4.6. Conceptual classification abstracts information by grouping semantically related features to form new abstract entities. Formally, it is defined as a non-injective function $\kappa : \mathcal{D}^n \to \mathcal{D}^m (m, n \in \mathbb{N})$,

$$\begin{aligned}\kappa(s_1, s_2, \dots, s_n) = (\kappa_1(s_{1,1}, s_{1,2}, \dots, s_{1,h_1}), \kappa_2(s_{2,1}, s_{2,2}, \dots, s_{2,h_2}), \dots, \\ \kappa_m(s_{m,1}, s_{m,2}, \dots, s_{m,h_m}))\end{aligned} \tag{4.3}$$

with $\kappa_i : D^{h_i} \to D$ and $h_i \in \{1, \dots, n\}$, where $i \in \{1, \dots, m\}$.

Conceptual classification subsumes the other two abstraction concepts, as the following theorems show:

Theorem 4.1. *Given a conceptual classification κ as defined in Def. 4.6, if all κ_i have the form $\kappa_i : \mathcal{D} \to \mathcal{D}$ and $m = n$, κ is a coarsening.*

Theorem 4.2. *Given a conceptual classification κ as defined in Def. 4.6, if all κ_i have the form $\kappa_i(s_j) = s_j$, $i \leq j$, $m < n$, and $\kappa_i = \kappa_j \Rightarrow i = j$, κ is an aspectualization.*

Both theorems are left without a proof, because the theorems follow directly from the definitions.

Example 4.4. Data gathered from a laser range finder comes as a vector of distance values and angles to obstacles in the local surroundings, which can be represented

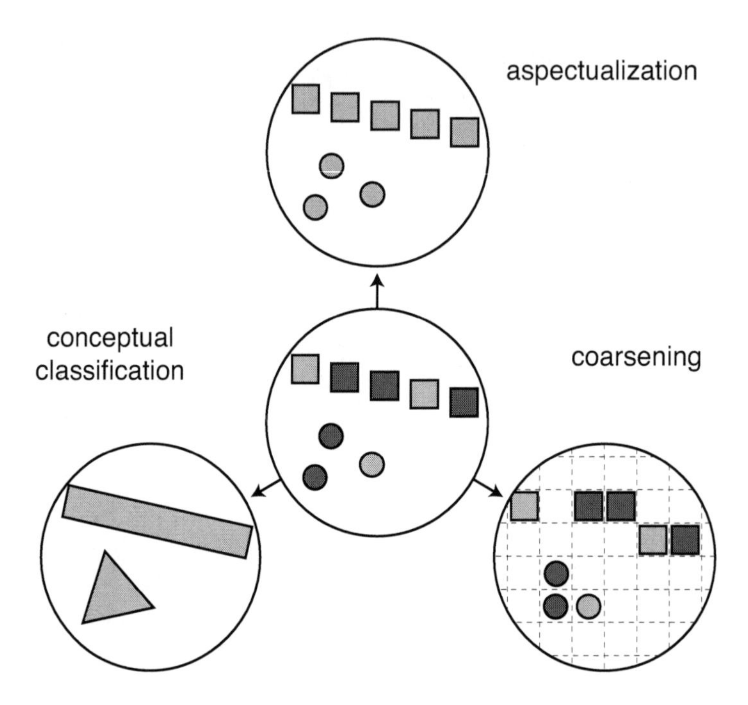

Fig. 4.6: Illustration of the three abstraction principles, aspectualization, coarsening, and conceptual classification, applied to the same representation

as 2-D points in a relative coordinate frame around the sensor. Abstraction of these points to line segments by the means of a line detection algorithm, as, for example, described in Gutmann et al. (2001), is a conceptual classification.

Figure 4.5 represents conceptual classification, and Fig. 4.6 summarizes the presented three facets of abstraction.

4.2.5 Related Work on Abstraction

Performing abstraction is a fundamental ability of intelligent agents and different facets of abstraction have thus been issued in related work, addressing various scientific fields and considering a rich diversity of tasks.

The insight that abstraction can be divided into different categories has been mentioned before. Stell and Worboys (1999) make a distinction between what they call "selection" and "amalgamation" and formalize these concepts for graph structures. The definitions of aspectualization and coarsening given in this chapter correspond to selection and amalgamation, which Stell and Worboys describe as being "conceptually distinct" types of generalization.

Bertel et al. (2004) also differentiate between different facets of abstraction ("aspectualization versus specificity," "aspectualization versus concreteness," and "aspectualization versus integration"), but without giving an exact definition. "Aspectualization versus specificity" corresponds to the definition of aspectualization given in Def. 4.2, and "aspectualization versus concreteness"corresponds to coarsening (Def. 4.4). However, the definition of aspectualization given in this work is tighter than the one of Bertel et al.—according to them, aspectualization is "the reduction of problem complexity through the reduction of the number of feature dimensions." In the framework presented in this chapter, the result of aspectualization is also a reduction of feature dimensions, but it is also explicitly required that all the values of the remaining feature dimensions be unchanged.

The notion of *schematization*, which Leonard Talmy describes as "a process that involves the systematic selection of certain aspects of a referent scene to represent the whole disregarding the remaining aspects" (Talmy, 1983) is tightly connected to the definition of aspectualization in this chapter. If we assume the referent scene to be aspectualizable according to Def. 4.3, then the process mentioned by Talmy is aspectualization as defined in this chapter.

Annette Herskovits defines the term schematization in the context of linguistics as consisting of three different processes, namely abstraction, idealization, and selection (Herskovits, 1998). Abstraction and selection correspond to the definition of aspectualization (Def. 4.2), while Herskovits's definition of idealization is a coarsening according to Def. 4.4.

4.3 Abstraction and Representation

Aspectualization and coarsening describe two very different processes: The former reduces the number of features of the input, the latter reduces the variety of instances for every single feature. While aspectualization necessarily reduces the dimensionality of a representation, coarsening preserves dimensionality. However, we will see in the following that, depending on the representation, both facets of abstraction are equally expressive.

Depending on the representation of the feature vector, coarsening can produce a result that is equivalent to an aspectualization: Let one or more mappings κ_{c_i} in a coarsening be defined as mappings to a single constant value, $\kappa_i = z$, $z \in \mathcal{D}$. Assume all other mappings κ_{c_i} to be the identity function. Now, consider an aspectualization that retains exactly the components not mapped to single constant values z_i by the coarsening. The target space $\mathcal{T}$ created by this aspectualization results in a linear subspace of $\mathcal{S}$ that is a projection from the resulting space of the coarsening. In other words, the aspectualization has a canonical embedding in the result of the coarsening; and we can find a bijection that maps one into the other.

The existence of this process can be formalized as a theorem:

Theorem 4.3. *If κ_a is an aspectualization, then there exists a bijection φ_1 and a coarsening κ_c such that the following diagram commutes:*

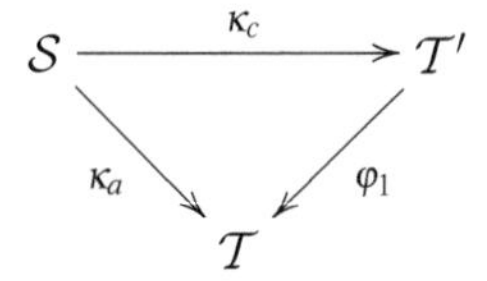

Proof. Following Def. 4.2, aspectualization conserves those features of $(s_1,\ldots,s_n)$ listed in an *aspectualization index vector* $I^\kappa = (i_1,\ldots,i_k)$, $I^\kappa \in \mathbb{N}^k$. Then, the *inverse aspectualization index vector* $I'^\kappa \in \mathbb{N}^n$ denotes the original position of the remaining features in the source feature vector:

$$I'^\kappa = (i'_1,\ldots,i'_n), \quad i'_j = \begin{cases} j|I^\kappa_j = i & \text{if } \exists j : I^\kappa_j = i \\ 0 & \text{else} \end{cases}. \tag{4.4}$$

Now we choose $\kappa_c : \mathcal{S} \to \mathcal{T}'$, $\kappa_c(s_1,\ldots,s_n) = \kappa_{c_1}(s_1),\ldots,\kappa_{c_n}(s_n)$ with

$$\kappa_{c_i}(s_i) = \begin{cases} s_i & \text{if } i \in I^\kappa \\ z & \text{else} \end{cases} \tag{4.5}$$

and a constant $z \in \mathcal{D}$. Furthermore, we choose $\varphi_1 : \mathcal{T}' \to \mathcal{T}$, $\varphi_1(t'_1,\ldots,t'_n) = (t'_{i_1},\ldots,t'_{i_k})$ with $i_j \in I^\kappa$ and $\varphi_1^{-1} : \mathcal{T} \to \mathcal{T}'$, $\varphi_1^{-1} = (\varphi_{1_1}^{-1},\ldots,\varphi_{1_n}^{-1})$,

$$\varphi_{1_i}^{-1}(t_1,\ldots,t_k) = \begin{cases} t_{I'^\kappa_i} & \text{if } I'^\kappa_i \neq 0 \\ z & \text{else} \end{cases}. \tag{4.6}$$

□

An example illustrates Theorem 4.3:

Example 4.5. As in Example 4.1, an oriented line segment in the plane is represented as a point $(x,y) \in \mathbb{R}^2$, a direction $\theta \in [0,2\pi]$, and a length $l \in \mathbb{R}$: $s = (x,y,\theta,l)$. κ_a is defined as in Example 4.1. The function $\kappa_c(x,y,\theta,l) = (x,y,1,l)$ is a coarsening, and κ_c delivers essentially the same result as an aspectualization:

$$\{\kappa_c(x,y,\theta,l)\} = \{(x,y,1,l)\} \cong \{(x,y,l)\} = \{\kappa_a(x,y,\theta,l)\}. \tag{4.7}$$

There exists a trivial bijection between both results.

Conversely, it is also possible to define a bijection that allows for expressing a coarsening by an aspectualization: If $\mathcal{S}$ is a mathematical structure, for example, a group, then this bijection even maintains the structure of $\mathcal{S}$, that is, it is an isomorphism:

Theorem 4.4. *If κ_c is a coarsening on a group, for example, $(\mathcal{S},+)$, then there exists an isomorphism φ_2 and an aspectualization κ_a such that the following diagram commutes:*

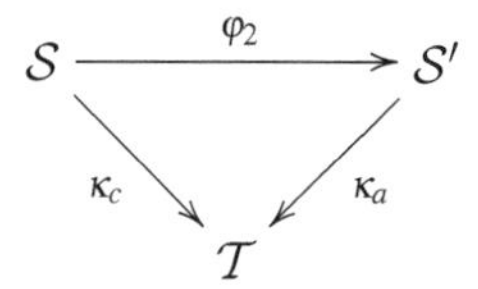

Proof. Choose $\varphi_2(s) = (s + \kappa_c(s), \kappa_c(s))$, $\varphi_2^{-1}(t_1,t_2) = t_1 + (-t_2)$ and $\kappa_a(t_1,t_2) = t_2$, and define $(\mathcal{S}',\oplus)$ with $\mathcal{S}' = \text{Image}(\varphi_2)$ and $t \oplus u = \varphi_2\left(\varphi_2^{-1}(t) + \varphi_2^{-1}(u)\right)$ for each $t,u \in \mathcal{S}'$. Again, checking that $(\mathcal{S}',\oplus)$ is a group and φ_2 is a homomorphism is straightforward. □

Again, an example illustrates the theorem:

Example 4.6. Coordinates $(x,y) \in \mathbb{R}^2$ can be bijectively mapped to a representation $(\lfloor x \rfloor, x - \lfloor x \rfloor, \lfloor y \rfloor, y - \lfloor y \rfloor)$ which features decimal places separately. The function $\kappa(x,x',y,y') = (x,y)$ is an aspectualization with the same result as the coarsening in Example 4.3.

Note that Theorem 4.4 does not add redundancy to the representation. If we would allow for introducing redundancy we could bijectively create new representations by concatenating s and an arbitrary abstraction $\kappa(s)$ with the effect that any abstraction, including conceptional classification, can always be achieved by an aspectualization from this representation. Therefore, we do not regard this kind of redundancy here.

Not every representation allows for coarsening, as the following example shows:

Example 4.7. Commercial rounding is defined by a function $f : \mathbb{R}_0^+ \rightarrow \mathbb{R}_0^+$, $f(x) = \lfloor x + 0.5 \rfloor$. f is a coarsening. If, similarly to Example 4.6, $x \in \mathbb{R}$ is represented as $(\lfloor x \rfloor, x - \lfloor x \rfloor)$, then commercial rounding can be expressed neither by aspectualization (because the representation is not aspectualizable with regard to this rounding) nor by coarsening (because the abstraction function operates on both components x and $x - \lfloor x \rfloor$ of the feature vector, which contradicts Def. 4.4). So even if commercial rounding reduces the number of instances in half of the components, the example above cannot be expressed as a coarsening under this representation following Def. 4.4. It then must be seen as a conceptual classification, which is the most general of the three facets of abstraction presented here.

Stell and Worboys (1999), among others, describe aspectualization and coarsening (or selection and amalgamation as they call it), as being "conceptually distinct." However, with regard to Theorems 4.3 and 4.4, this cannot be agreed on. The theorems show that, on the one hand, the result of an aspectualization can be achieved by a bijective translation of a coarsening and, on the other hand, there exists a bijection to transform any representation in a way that the result of a coarsening can be achieved by an aspectualization. The distinction between both facets of abstraction only applies to the *process* of abstraction and not the *result*. Different abstraction paradigms, even if describing distinct processes, can have the same effect: Applicability of a specific abstraction principle relies heavily on the given representation, and usually different types of abstractions can be used to achieve the same result. Thus, the choice of an abstraction paradigm is tightly coupled with the choice of the state space representation: It is always possible to find a bijection that transforms the representation in a way that we can express a coarsening by an aspectualization—and vice versa.

The following section argues for an action-centered view for choosing appropriate representations for the problem at hand.

4.4 Abstraction in Agent Control Processes

Abstraction, as defined above, is not an arbitrary reduction of information, but comes with a particular purpose. It is applied to facilitate solving a specific problem, and the concrete choice of abstraction is implied by the approach to master the task.

4.4.1 An Action-Centered View on Abstraction

To use spatial abstraction in the context of agent control tasks, three goals have to be reached:

1. significantly reducing the size of the state space the agent is operating in;
2. eliminating unnecessary details in the state representation;
3. subsuming similar states to unique concepts.

The first goal, reducing state space size, is a mandatory consequence of the latter two, which must be seen in the context of action selection: The question of whether a detail is "unnecessary" or whether two states are "similar" depends on the task of the agent:

- A detail is considered *unnecessary* if its existence does not affect the action selection of the agent.
- Two states are considered to be *similar* if the agent is expected to select the same action in any of the states.

Both points describe graded concepts: A detail may not be completely unnecessary, but may be without importance for action selection in a varying number of cases. Analogously, the amount of similarity of states could be described by regarding how often the difference leads to a different action selection.

This action-centered view expands classical definitions of similarity, as is for example given in Roberts (1973): Two states s and s' are indistinguishable (written

$s \sim s'$) if there is a mapping $f : \mathcal{S} \to \mathbb{R}$ and an $\varepsilon \in \mathbb{R}^+$ with $s \sim s' \Leftrightarrow |f(s) - f(s')| < \varepsilon$. Apart from the problem that this definition allows us to view any states as indistinguishable (with for example $f(s) \equiv 0$ and arbitrary ε), Roberts's concept is data-driven whereas our concept is action-driven in order to account for the task at hand. States may be very near with regard to a certain measure, but nevertheless require different actions to take in different contexts. Grid-based approaches achieved by coarsening, that is, state space coarsening as described in Sect. 3.2.1.1, bear the danger of not being able to provide an appropriate state separation due to missing resolution in critical areas of the state space. Furthermore, a "nearness" concept as presented by Roberts is again a matter of representation and may only be appropriate in homogeneous environments.

Abstraction is supposed to facilitate the agent's action selection process. If those details that are irrelevant for the choice of an action are eliminated, difficulty and processing time of action selection are reduced, and action selection strategies may be applicable to a broader range of scenarios.

When choosing an abstraction paradigm for a given data set, the result must be regarded in the context of *accessibility* of information. The goal of abstraction must be to enable easy access to the *relevant* information. Andre and Russell (2002) call the ability to decide only on relevant features a "critical aspect of intelligence." What piece of information is relevant depends on the task at hand: A computer-driven navigation control may require different concepts than a system interacting with a human being. Abstraction retains information that is relevant for a certain purpose. Therefore, it can never be regarded as purely data-driven, but requires a solid a priori concept of the problem to solve and, consequently, the actions to take.

4.4.2 Preserving the Optimal Policy

Our action-centered view on abstraction is also shared by the definition of *categorizability* given in Porta and Celaya (2005, see also Sect. 3.4.2). The authors call an environment categorizable if "a reduced fraction of the available inputs and actuators have to be considered at a time." Categorizability in this context is a graded concept, not a binary one. In a highly categorizable environment, we can achieve an abstraction that subsumes those states to identical representations that lead to an identical action selection. Put differently, we seek for an abstraction that takes decision boundaries into account (compare this to the approach of Reynolds, 2000, described on p. 35). Such an abstraction preserves the optimal policy π^* for the agent to reach the goal, so we call this abstraction *π^*-preserving*:

Definition 4.7. Let $\pi^*_{\mathcal{S}}$ and $\pi^*_{\mathcal{T}}$ be optimal policies for tasks operating on state spaces $\mathcal{S}$ and $\mathcal{T}$. An abstraction $\kappa : \mathcal{S} \to \mathcal{T}$ is called π^*-preserving, if for every $s \in \mathcal{S}$ it holds that $\pi^*_{\mathcal{S}}(s) = \pi^*_{\mathcal{T}}(\kappa(s))$.

In particular, this means that for all states that are mapped to the same abstract state the optimal action is identical, that is, for all $s_1, s_2 \in \mathcal{S}$ with $\kappa(s_1) = \kappa(s_2)$ it holds that $\pi^*_{\mathcal{S}}(s_1) = \pi^*_{\mathcal{S}}(s_2)$. In an MDP setting, this kind of abstraction is also known as *policy-irrelevant* abstraction (Jong and Stone, 2005) or π^**-irrelevant* abstraction.

In a framework presented by Li et al. (2006), policy-irrelevant abstraction is presented as the coarsest type amongst 5 types of state abstraction for MDPs. But even this coarse type of abstraction may be infeasible to achieve in most applications. However, the benefits of abstraction will balance out the fact that the resulting policy will be slightly suboptimal. So in most cases it is sufficient to find an abstraction that fulfills the criteria for π^*-preservation for a subset of $\mathcal{S}$. This kind of abstraction is called *partially* π^**-preserving*. Partial π^*-preservation is a graded concept as well. In finite state spaces, we can measure its magnitude, the π^**-preservation quota* of an abstraction κ:

$$q_{\pi^*}(\kappa) = \frac{|\{s \in \mathcal{S} | \pi^*_{\mathcal{S}}(s) = \pi^*_{\mathcal{T}}(\kappa(s))\}|}{|\mathcal{S}|}. \tag{4.8}$$

Obviously, it holds that $q_{\pi^*}(\kappa) \in [0,1]$. An abstraction κ with $q_{\pi^*}(\kappa) = 1$ is π^*-preserving. Unfortunately, value functions and, thus, the policy may be only partially defined, as unvisited states are omitted. Furthermore, specifying a concrete value for $q_{\pi^*}(\kappa)$ in continuous domains is generally infeasible.

π^*-preservation is tightly connected to the concept of *MDP homomorphisms* (Ravindran, 2004, see Sect. 3.5.3), but more general. MDP homomorphisms also require the exact conservation of the system dynamics (transition probabilities and rewards) more than just an identical action selection. Also, it relates to the concept of *safe abstractions* defined in a hierarchical learning architecture (Andre and Russell, 2002) that are π^*-preserving while also maintaining Q-values. An early work referring to safe abstractions, although without mentioning them explicitly, is Amarel's paper on reasoning about actions (Amarel, 1698).

4.4.3 Accessibility of the Representation

As described in Sect. 4.3, different abstraction paradigms can be used to achieve the same effect, given an appropriate state space representation. From the viewpoint of accessibility the use of aspectualizable representations is to be preferred, as relevant aspects are clearly separated and easy to access. The aspectualized features form a reduced representation of the aspect on their own, and (partial) problems that rely on this aspect become manageable. This reduced representation does not need to be computed, but is immediately available after a selection process of features.

From a cognitive perspective, Bertel et al. stress that aspectualization is a selection rather than an omission. Only a limited number of feature dimensions is regarded, while the arbitrarily large amount of remaining features is silently ignored and not processed explicitly. Thus, aspectualization is a computationally simple process. The benefits of aspectualization lie in "a reduced processing load ..., a more focused context, and stronger structural analogies between problem, problem solving, and problem representation" (Bertel et al., 2004). Even if this has been stated in the context of architectural design, it also holds for computational data processing. Aspectualizability facilitates knowledge extraction, and furthermore, knowledge transfer. How aspectualizability relates to the questions of generalization and transfer learning is the topic of Chap. 5.

If nothing else, aspectualizable representations are also easy to comprehend for humans, as the relevant aspects can be read from the abstracted feature vector without any further interpretation of data. This becomes helpful for interaction between humans and machines; however, this is not discussed in greater detail within this book.

If no aspectualizable representation is present, it is beneficial to create one. One possibility is to transform coarsenings into aspectualizations (Theorem 4.4). Another possibility is to use abstraction itself; in particular, conceptual classification is a powerful means to create new entities in an aspectualizable manner. So abstraction helps to create abstract representations that allow for distinguishing between different aspects by using aspectualization.

4.5 Spatial Abstraction in Reinforcement Learning

In Sect. 4.4.1 we have seen that abstraction can be used to describe concepts, and that those concepts are used to trigger the right action for a certain purpose. The more the states that can be subsumed to one concept with one consequent action, the smaller the state space becomes. If this is consequently thought to the end, one can conclude that an optimal action-centered abstraction is a function that maps all states that trigger the same action to a singular concept. This optimal abstraction is then π^*, the optimal policy for the problem: $\pi^*(s) = \text{argmax}_a Q^*(s,a)$—exactly the desired result of the learning process. From this point of view, Q abstracts states to action concepts.

4.5.1 An Architecture for Spatial Abstraction in Reinforcement Learning

The final abstraction π^* emerges from an autonomous learning process and builds a partition over the state space, and it is this selection of states into categories (which requires to visit all the states repeatedly) that is the time-consuming part. If the state space is smaller, the number of states to categorize decreases, and so does the learning time—if the decrease happens in a way that preserves the system dynamics; that is, the subsumption of states must be done such that it summarizes states that request the invocation of the same action in order to reach the goal of the task.

In this work we achieve this subsumption by applying abstraction techniques directly to the incoming data from a sensory system. The state space the reinforcement learning algorithm operates on is an abstraction of the incoming sensory data. In-

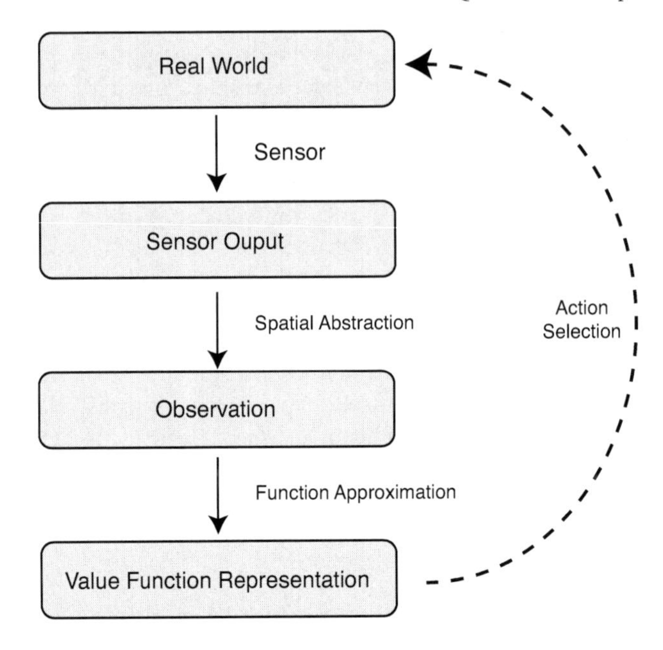

Fig. 4.7: Abstraction is applied directly to the data coming from the sensory system

stead of operating on the state space $\mathcal{S}$ acquired by the sensory system, the agent is assumed to operate on a set of newly created entities which we call *observations*.

Figure 4.7 shows the proposed architecture for a learning system: The sensors of the autonomous agent perceive a state s from the real (or simulated) world and create the state space $\mathcal{S}$. Instead of now storing s in a value function representation (which in continuous state spaces requires some sort of function approximation), another processing step is added. We use an abstraction κ to map $\mathcal{S}$ to an observation space $\mathcal{O}$ and transfer the state s to an observation $\omega = \psi(s) = \kappa(s)$. If needed, function approximation is then applied to $\mathcal{O}$, and actions can be derived from the stored value function V (or Q).

Usually, the input an artificial agent receives comes from its sensory system. Even this data is already an abstraction of the real world: It reduces the rich information of the surrounding world to a camera image or a number of distance values. Depending on the kind of sensor, this abstraction may be stronger or weaker, but no sensor, including the human senses, can ever provide a complete picture of the surrounding world. Anyway, this information is enough for humans and artificial agents to develop successful problem solving strategies in the space they operate in. The aim must be to intensify this abstraction to achieve an observation space that allows for efficient problem solving without lowering the quality of the solution. In other words, the aim is to find a π^*-preserving abstraction.

As stipulated in Sect. 4.1, the invocation of spatial abstraction at this point does not change the learning algorithm. Thus, this approach can easily be combined with existing approaches addressing the same problem. Spatial abstraction is not meant to be a replacement for any algorithm for efficient learning. It is meant to be a means to provide an observation space that these algorithms can cope with. This book explicates the effectiveness of spatial abstraction by showing that it enables us to solve complex control problems even when using standard reinforcement learning techniques.

4.5.2 From MDPs to POMDPs

This book concentrates on problems that can be formulated as a Markov decision process $\langle \mathcal{S}, \mathcal{A}, T, R \rangle$ (see Sect. 2.3). Reinforcement learning techniques such as Q-learning are a means to solve MDPs, where the agent operates on a state space $\mathcal{S}$. Section 4.5.1 proposed an architecture applying a spatial abstraction as an observation function $\psi(s) = \kappa(s)$ directly to $\mathcal{S}$ to achieve an observation space $\mathcal{O}$.

Due to the abstraction, subsets of $\mathcal{S}$ collapse into single elements of $\mathcal{O}$, so the original states are no longer distinguishable. That is, the problem becomes partially observable, and our problem is not an MDP anymore, but a POMDP (Sect. 2.3.3). Operating on $\mathcal{O}$ instead of $\mathcal{S}$ makes the system lose the Markov property. Thus, the solution techniques for MDPs are not applicable anymore to obtain a guaranteed solution to the problem. However, solutions to a POMDP are very hard to obtain. Introducing this kind of complexity would be to oppose to the goal of easing the learning process to make it more efficient.

A strikingly simple idea is to ignore the partial observability property and apply solution techniques for MDPs to POMDPs. That is, we use ψ to replace $\mathcal{S}$ with $\mathcal{O}$ and solve the resulting problem as if it were an MDP.

Due to the partial observability, the resulting policy will in almost every case not be optimal. One cannot expect to use abstraction and lower the granularity of the state space without a price. A quality loss must always be taken into consideration when throwing away information. If we want to increase the learning performance with regard to speed, facilitate of use, robustness, and so on, a slightly worse solution is acceptable in the majority of cases. Nevertheless, we must aim at finding abstraction techniques that keep this loss as small as possible.

The quality of a solution using this approach critically depends on how tightly the system dynamics of POMDP and the underlying MDP are coupled. So again, the success of abstraction hinges on its π^*-preservation quota: the higher it is, the better is the quality of the found solution.

It must be noted that even a π^*-preserving abstraction can lead to suboptimal policies with Q-learning, and even convergence cannot be guaranteed (Li et al., 2006). Choosing a finer type of abstraction in the frame work of Li et al. is no alternative, though: Those types of abstraction require particular knowledge of the optimal Q-function Q^*. When designing an abstract state space for a given problem, this requirement is crucial: Q^* is supposed to be result of our learning efforts and is not available in the beginning. Even more, in non-trivial learning tasks, the Q-values of the optimal Q-function cannot even be estimated a priori. In contrast to that, assumptions on what might be the best possible action in a class of states are much easier to make. They originate from a certain domain knowledge about the characteristics of a task that is often given when designing the state space representation for a certain problem (see Sect. 4.5.5). In many cases, the system designer has a clear idea of which types of states should be collapsed to achieve a high π^*-preservation quota. Thus, π^*-preservationis a reasonable concept to aim at when setting up the learning task.

4.5.3 Temporally Extended Actions

Applying abstraction to a state space means collapsing several states $s_1, s_2, \dots$ into an abstract state or observation ω, and learning a value function $V(\omega)$. The value function denotes the overall expected reward when following the current policy (see Sect. 2.3.2). Under state space abstraction, these semantics are most likely to be broken. $V(\omega)$ denotes an average over the expected rewards of the underlying states, and this $V(\omega)$ generally differs from $V(s_1)$. Even if we have a π^*-preserving abstraction, this might lead to suboptimal action selection in the overall task, because the value of the abstract state reached by executing the optimal action might be lower than that of another neighboring state.

Fortunately, this does not happen when using the model-free Q-learning algorithm (see Sect. 2.5.3). Here, a Q-value is stored for every state-action pair (s,a). Thus, a greedy action selection does not regard neighboring states, but only relies on the Q-value $Q(s,a)$. In a π^*-preserving abstraction, $\text{argmax}_a Q(s,a)$ is the same for any s in the same abstract state, so the agent will follow identical (optimal) actions a while perceiving the same observation $\kappa(s)$.

In this case, there is no need for a decision at every time step, because we can expect it to be the same. Thus, *temporarily extended actions* can be added to the learning process: As long as the observation does not change, no new decision is necessary, that is, the action $a^* = \text{argmax}_a Q(\kappa(s), a)$ will be executed until a state s' is reached with $\kappa(s') \neq \kappa(s)$. This corresponds to the approach taken in the TTree algorithm (Uther and Veloso, 2003, see Sect. 3.4.3). The effect is a decrease in state transitions and decisions, leading to an increased learning speed.

Extended actions also make sense from a cognitive point of view: While the observation of the world on which the action selection is based does not change, what reason could there be to revoke this decision? A new decision is only necessary if the agent notices a difference in its observation of the environment.

4.5.4 Criteria for Efficient Abstraction

In the following, let us make some claims on how a spatial abstraction κ should be designed to constitute a suitable observation function $\psi : \mathcal{S} \to \mathcal{O}$, $\psi(s) = \kappa(s)$, in the reinforcement learning framework. In particular, it is desirable to achieve the following properties:

Significant Size Reduction. The observation space has to be significantly smaller than the original state space $\mathcal{S}$. According to Def. 4.2, even a subsumption of two states to one observation would be enough to achieve an abstraction, but of course this would not show a significant effect on learning performance.

Obtaining a Discrete State Space. A complex control problem usually operates in a continuous state space, which can become cumbersome because of the need to apply function approximation to the value function (see Sect. 3.2). In the best case, the observation function ψ will map to a discrete observation space $\mathcal{O}$ with a finite number of states. A discrete observation space allows for storing the value function in an exact tabular representation. Several efficient RL algorithms, for example, SPITI (Degris et al., 2006), only operate on discrete state spaces. So obtaining a discrete state space widens the range of possible learning algorithms to apply to the task.

High Generalization. To allow for efficient reuse of learned knowledge it is desirable to collapse a large number of states into single representatives. In an optimal (but theoretical) case, all states triggering the same behavior should be subsumed into one single observation.

High π^*-Preservation Quota. Conversely, we want to preserve the optimal behavior also in the generalized observations. In other words, a high π^*-preservation quota is desirable. How this can be achieved is examined later on.

High Accessibility Through Aspectualizability. Aspects that are important for action selection should be explicitly modeled. That implies that the resulting observation has to be easily accessible. For example, Chap. 5 explicates that for knowledge transfer this plays a relevant role. Thus aspectualizability of the observation with regard to aspects relevant for action selection is desired.

4.5.5 The Role of Domain Knowledge

When choosing an abstraction paradigm for a certain task, we aim at finding an abstraction with a high π^*-preservation quota. Unfortunately, this cannot be ensured at the beginning of the learning process, and knowledge about the system dynamics is foremost something that is achieved during learning. To find an appropriate abstraction, basic knowledge about the domain we operate in is required. As stated above, an action-centered view on abstraction requires a solid a priori concept.

When choosing to solve a problem with autonomous learning techniques, the hope is to achieve a solution that goes beyond the known concepts and understanding of the underlying system. However, in practical use, one cannot abstract from any domain knowledge if the problem grows complex. Several questions regarding the domain of the task have to be answered before designing a learning system. For example, the learning algorithm and its parameters, such as learning rate and exploration rate, have to be chosen carefully, depending on the problem. Also the choice of function approximation techniques usually requires solid expert knowledge in the domain of the task and the learning techniques.

Specifically, knowledge about structural properties of the state space is important to choose an abstraction that takes this structure into consideration and represents it explicitly. For many applications we have solid knowledge of structuring entities. Indoor environments, for example, are structured by walls and doors. This is a simple insight, but for an autonomous system it is hard to learn this from scratch. Furthermore, the agent has no information on how to cope with this structure. However, consideration of such facts can help us build reasonable concepts, and spatial abstraction can transfer this kind of background knowledge into an explicit representation and, thus, facilitate the detection of structure within a state space.

4.6 A Qualitative Approach to Spatial Abstraction

As pointed out in Sect. 4.4.1, the coupling of actions with the abstract representation is essential for efficient agent control processes. Different consequences of an observation define different *qualities* of the situation, and thus states can be classified according to these qualities. Following that, so-called *qualitative representations* are an appropriate choice when using abstraction in agent control tasks, because they are designed to model what is essential for a task at hand.

4.6.1 Qualitative Spatial Representations

Qualitative representations do not rely on the precision of information. They abstract from details in the world and concentrate on describing entities within the world and their relations to each other. Qualitative representations are based on concepts such as "left of" or "behind," or "far" and "near" rather than on precise measurements. A fundamental property of a qualitative representation is to denote properties of a continuous world with a finite set of discrete symbols (Cohn and Hazarika, 2001).

Qualitative *spatial* representations, that is, representations that are grounded on the "real" 3-D space humans operate in or its 2-D projections, are important concepts in spatial cognition research. They have especially shown growing importance over the last few years and have been successfully integrated in many applications. As many control processes operate within environments humans also operate in, qualitative spatial representations are excellent candidates for modeling agent control tasks. For example, they have been included in Benjamin Kuipers's robot navigation architecture *spatial semantic hierarchy* (Kuipers, 2000), as well as in the already mentioned multi-robot application described by Sierra et al. (2001, see Sect. 3.4.4), or in robot soccer (Schiffer et al., 2006).

When a certain set of operations is defined on a qualitative representation, we speak of a *qualitative calculus* (Ligozat and Renz, 2004). Qualitative spatial calculi allow for expressive reasoning about qualitative concepts, which establishes the research field of *qualitative spatial reasoning* (QSR). There are reasoning engines especially dedicated to qualitative spatial reasoning, for example, SparQ (Wallgrün et al., 2007; Wolter, 2009) and GQR (Gantner et al., 2008).

However, QSR still plays a minor role in agent control processes. Examples of the few approaches to apply QSR to agent control are the robot path planning system by Escrig and Toledo (2000), which uses the double-cross calculus (Freksa, 1992; Freksa and Zimmermann, 1993) and the SailAway application for rule-compliant control of autonomous vessels (Dylla et al., 2007), using the oriented point reasoning algebra $\mathcal{OPRA}_m$ (Moratz, 2006; Moratz et al., 2005). In this book, we concentrate on qualitative representations, without taking reasoning capabilities into account.

4.6.2 Qualitative State Space Abstraction in Agent Control Tasks

Abstraction, as introduced in this chapter, is defined as a non-injective function. That means that several entities in the original state space can share one abstract representation, a fact that, according to Galton and Meathrel (1999), is "in the nature of a qualitative representation system." Thus, qualitative representations are well suited for representing the abstractions desired in control processes.

Qualitative representations distinguish between different descriptions based on their importance for a certain task to be solved. Freksa (1992) points out that "reasoning based on qualitative information aims at restricting knowledge processing to that part of the information which is likely to be relevant in the decision process: the information which already makes a difference." That is what Forbus (1984) calls the *relevance principle*: The distinctions a representation makes have to be relevant to the kind of reasoning being performed. As qualitative representations follow the relevance principle, they are well suited to model the state space's *structure*.

In the context of reinforcement learning, the relevant places are at decision boundaries. It is especially important at places where the value function is inhomogeneous, that is, at places where it shows significantly differing values for near states. Adaptive state space abstraction techniques (Sect. 3.4.1), for example, implement the idea to increase the resolution of the state space representation at these places by applying coarsening to the state space.

As Cohn and Hazarika (2001) formulate, qualitative reasoning even *refuses* to differentiate between details unless there is sufficient evidence that it is really needed. Identical representations for states that require identical decisions for action selection are fundamental for a qualitative representation, because qualitative knowledge is the aspect of knowledge which critically influences decisions (Freksa and Röhrig, 1993). As a consequence, this allows for generalization within a learning task. So a reasonable qualitative representation matches the "high generalization" and "high π^*-preservation quota" criteria for efficient abstraction worked out in Sect. 4.5.4. Also, the criterion "discrete space space" is fulfilled naturally when using this type of representation. The request for a "significant reduction of state space size" is also granted. Thus, qualitative spatial representations are best suited for the purpose of efficient abstraction for reinforcement learning.

4.6.3 Qualitative Representations and Aspectualization

The last remaining criterion from Sect. 4.5.4 is "high accessibility through aspectualizability." Qualitative representations describe entities and their relations, and each of them can be considered to be a relevant *aspect* for solving the task at hand if we assume the representation to be suitable. Freksa and Röhrig (1993) explicitly state that qualitative representations describe "aspects of knowledge." So the problem reduces to identifying relevant aspects, which has been taken care of within the design process of the representation for the given task. What is particularly relevant, that is, which aspects are to be explicated, depends on the task at hand, but in any case the structure of a domain is relevant. A representation based on the structure of the state space bears spatial knowledge about the environment, which is advantageous because spatial constraints are internally provided by the input representation and do not need to be acquired separately.

If the structure can be explicated in an aspectualizable manner, the request for high accessibility is also fulfilled. However, the aspects need to be extracted from rich metrical sensory information: For example, the distinction between whether an object is to the left or to the right of a robot must be calculated from the distance values of a laser range finder or extracted from camera images and represented adequately. As stated before, aspectualizability of a feature vector can be achieved by applying conceptual classification on the raw data of sensory perception in order to group relevant conceptual entities into single aspects. Choosing and representing relevant features also tackles the question of how to achieve non-local generalization. This is addressed in the next chapter.

Qualitative spatial representations offer additional benefits: As they are usually based on human spatial concepts, they offer a straightforward symbol grounding. This can, for example, support easier interaction with humans. In the context of autonomous learning, this can become important for incorporating human domain knowledge as background information into the learning process. However, this is beyond the scope of this book.

4.7 Summary

In this chapter, we took a deeper look at abstraction. In particular, abstraction of the state space has been identified as a suitable means to tackle the challenges of reinforcement learning in complex state spaces.

A formal definition of abstraction has been given that distinguishes between its three different facets: aspectualization explicates features by ignoring the rest, coarsening lowers the granularity of the state space, and conceptual classification builds new entities by grouping semantically related features. With the formal definitions of abstraction principles the various terms used to describe abstraction can very well be classified and related, and the formalisms allow for insights into the nature of abstraction. In particular, we have seen that aspectualization and coarsening, which are usually regarded as conceptually distinct, can be used to achieve the same effect if we allow for bijections in source and target space. Thus, only the process of abstraction is conceptually distinct, but not the result. Applicability of abstraction depends strongly on the chosen representations.

To successfully apply abstraction in agent control processes an action-centered view on abstraction that concentrates on the decisions being drawn at certain states becomes essential. Starting from that, π^*-preservation has been defined as a central concept for meaningful spatial abstraction. The success of autonomous learning depends heavily on the description of the world. Because accessibility facilitates knowledge extraction, aspectualizable representations are the key for efficient learning. Aspectualizable feature vectors can be created by using abstraction, in particular, conceptual classification.

Abstraction should be applied directly to the incoming data such that sophisticated learning mechanisms still can be applied. Even if this breaks the Markov assumption, learning can lead to a near-optimal policy if the abstraction is meaningful. Therefore, five criteria for efficient abstraction have been derived: significant size reduction, discreteness of state space, high generalization, high π^*-preservation, and high accessibility through aspectualizability. These criteria can most satisfactorily be matched by the use of qualitative representations, because in the propagated action-centered view they stress the details that already make a difference. Based on a solid a priori concept, aspectualizable qualitative abstraction enables for efficient application of reinforcement learning in complex state spaces.

Chapter 5
Generalization and Transfer Learning with Qualitative Spatial Abstraction

In this chapter we will investigate the properties of observation space representations achieved by qualitative abstraction with respect to generalization and transfer learning capabilities. Section 5.1 describes the importance of structural similarity for knowledge reuse. Two different aspects of behavior of an agent in a control task are identified in Sect. 5.2 and brought together with the abstraction paradigm of aspectualization to achieve so-called structure space aspectualizable state spaces. Based on this concept, two different methods for transfer learning are introduced in Sects. 5.3 and 5.4, namely task space tile coding (TSTC) and a posteriori structure space transfer (APSST). The approaches are compared and discussed in Sect. 5.5. SITSA, an approach to manipulate the observations of an agent based on structural knowledge gained in previous tasks, is presented in Sect. 5.6, before the chapter closes with a summary.

5.1 Reusing Knowledge in Learning Tasks

As described in Chap. 3, knowledge reuse is a critical issue for empowering an agent to successfully learn a complex control task, especially in large and continuous state spaces. In particular, non-local generalization and cross-domain transfer learning are important aspects of successful learning and, thus, deserve further investigation.

It has been already pointed out in Sect. 3.1.2 that for effective reuse of already acquired knowledge—that is, for achieving non-local generalization and cross-domain transfer—the representation of structure within the domain is the crucial point. In the following, the correlations of knowledge reuse, structure, and representation will be studied further.

L. Frommberger, *Qualitative Spatial Abstraction in Reinforcement Learning*,
Cognitive Technologies, DOI 10.1007/978-3-642-16590-0_5,

5.1.1 Structural Similarity

A key capability of an autonomous agent is to benefit from similarities in the space it is operating in. A slight variation in its position, for example, will most of the time not lead to any difference in the assessment of its state. If a robot is in a corridor and wants to follow it, it is not important whether it is exactly in the middle of the corridor or slightly to the left or to the right of the middle—the action selection in this situation should result in "go straight ahead". The positions are *similar* enough to make us not care about the differences.

Similar states do not necessarily need to be "close" to each other and indistinguishable (as for example proposed by Roberts (1973)). In Sect. 4.4.1, we defined similarity from an action-centered point of view: states are similar if they induce the same action.

Different tasks (or different situations or places within a task) are mostly not completely distinct. Often, they are in some way related, that is, they share some commonalities. Knowledge gained in one task may not necessarily be useless in another, and action selection to solve the tasks may partially lead to related actions of the autonomous agent. Because such commonalities induce a certain, recognizable behavior, we can regard them as a *structure* of the state space (or observation space). State spaces that share structures, problems, or tasks that are defined on them are called *structurally similar*.

Example 5.1. An office environment consists mainly of walls that form rooms and corridors. The behavior of agents operating in this environment is highly dependent on the walls, as they build the structure of the world, and actions such as "following the corridor" are directly affected by this structure. Any other office environment has walls, corridors and rooms as well—office environments constitute structurally similar state spaces. Tasks that take place in these environments—for example, finding the restrooms in a building—are structurally similar as well.

These structural similarities may be explicitly or implicitly represented in the feature vector. Of course, explicitly represented structural elements—at the best in an aspectualizable manner—are easier to access. For the agent, noticing structural similarities is an important step towards knowledge reuse.

5.1.2 Structural Similarity and Knowledge Transfer

Both generalization and transfer learning aim at making use of the similarities within or across tasks by using previously gained knowledge for the current learning process. Knowledge transfer of this kind is only possible in structurally similar spaces. Thus, detection of structural similarity allows for local and non-local generalization as well as cross-task transfer learning. When the agent is able to notice that a state it is in is structurally similar to a state of which it already gathered solid knowledge,

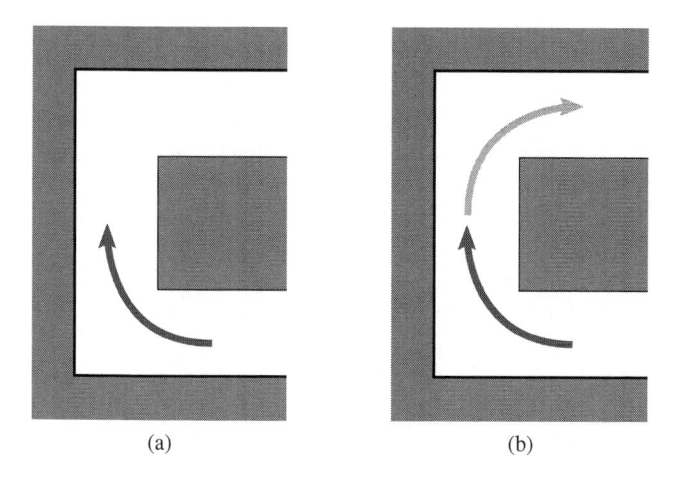

Fig. 5.1: Generalizing knowledge: When we know how to turn around the first corner (a), does this knowledge help us to turn around the second one (b)?

then learning at this state does not have to start from scratch: Structurally similar states require similar actions.

Figure 5.1 shows a corridor with two turns. Assume the robot has learned how to take the curve in Fig. 4.3a. Is it now also able to use this knowledge to turn around the second curve depicted in Fig. 4.3b? Does it notice it is in a structurally similar situation?

So the main questions to be answered are:

- When are states structurally similar? When are they not similar?
- To which part are they similar?
- How does the learning agent notice similarity?
- How is structural similarity represented in the feature vector?
- How do we access the knowledge to benefit from this similarity?

To answer these questions, we now investigate how knowledge reuse at structurally similar places can be realized.

5.2 Aspectualizable State Spaces

In this section, we now focus on the actions the agent performs according to the policy it has or is about to learn. The choice of actions defines a behavior of the agent. How this behavior is influenced by the structure of the domain is investigated in the following.

5.2.1 A Distinction Between Different Aspects of Problems

Let us say we have two structurally similar tasks such as finding restrooms in two different environments, as mentioned in Example 5.1. The sequence of actions the agent performs will generally be different if the two state spaces are not identical. From its starting position, it will perform movements that are dependent on the structural elements such as corridors, but also depend on where the agent starts and where the goal states—the restrooms—are located.

However, in certain parts of the two environments, we will experience a locally identical behavior of the moving agent. If we assume an egocentric view, the actions to perform to follow a corridor or move around a turn are the same, regardless of where the corridor is (it can be in the same environment or in a different one). So the agent's behavior depends on two separate aspects—one that is tightly coupled with the goal to be reached and one that is influenced by certain structural elements of the state spaces.

So we distinguish two aspects within the behavior of the agent, *goal-directed behavior* and *generally sensible behavior.*

1. *Goal-directed behavior* towards a certain goal state depends highly on the specific task to be solved. In general, resulting actions at distinct states are different for different goal states or in different environments. Goal-directed behavior is specific to the task at hand, and so the corresponding knowledge about the behavior is *task-specific knowledge*.
2. *Generally sensible behavior* regards the structure of the environment, so it is more or less the same in structurally similar state spaces. It does not depend on a goal to reach, but on structural characteristics in the environment that demand a certain behavior. Generally sensible behavior is independent of the task at hand, and so the corresponding knowledge about the behavior is *task-independent knowledge*.

Example 5.2. Generally sensible behavior of an agent operating in the office environments given in Example 5.1 includes not colliding with walls, turning around corners smoothly, proceeding through corridors in a straight line, and avoiding superfluous movements. In contrast to this route-conforming behavior, any decisions about which route to select at intersections refer to goal-directed behavior.

5.2.2 Using Goal-Directed and Generally Sensible Behavior for Knowledge Transfer

Goal-directed and generally sensible behavior are not completely independent. On the one hand, reaching a target location requires some sort of generally sensible behavior. Knowledge of generally sensible navigation behavior is a good foundation for developing goal-oriented strategies. For example, it helps in exploring the environment. There must be a generally sensible behavior in every successful goal-oriented strategy; otherwise the goal state would not be reached. Therefore, generally sensible behavior is a good candidate for knowledge transfer, because it is existent and useful in any structurally similar task.

On the other hand, a behavior that works successfully in any task still relates to a certain purpose. In a robot navigation task, for instance, driving forward into a corridor is a generally sensible behavior—if the goal of the agent is to drive to a certain location. If the goal is not to collide, then doing nothing would probably be a better option. Thus, when transferring generally sensible behavior, the type of the new task must conform with the original one.

5.2.3 Structure Space and Task Space

Generally sensible behavior depends on structural characteristics of the environment. To take those characteristics into account, it is desirable that the structure of the state space be reflected in the state space representation. The structure is important for the agent's behavior, so we want it to be easily accessible. As has been argued in Sect. 4.4.2, it is especially aspectualization that provides high accessibility.

A state space representation that is aspectualizable with regard to structural characteristics explicitly encodes the structure in a separate set of features. The corresponding feature dimensions span a feature space that we call *structure space*. The remaining feature dimensions do not explicitly represent structure and belong to goal-directed behavior for a particular task. Thus, this set of features spans *task space*.

In other words, the overall observation space $\mathcal{O}$ should be represented as a Cartesian product of task space $\mathcal{O}_T$ and structure space $\mathcal{O}_S$, that is, $\mathcal{O} = \mathcal{O}_T \times \mathcal{O}_S$. The desired observation function $\psi : \mathcal{S} \to \mathcal{O}$ then consists of two distinct partial observation functions. So ψ should have the form $\psi(s) = (\psi_T(s), \psi_S(s))$.

5.2.3.1 Structure Space Aspectualization

Both task space representation and structure space representation can be derived from an observation $\omega \in \mathcal{O}$ by an aspectualization. Thus, an observation space of this form is called *structure space aspectualizable* or *task space aspectualizable*, respectively. Structure space aspectualizable observation spaces allow for easy reuse of generally sensible behavior. In particular, it is the careful design of observation function and observation space representation that enable transfer of structure space knowledge.

To formalize this aspectualization, we define two functions, $\mathrm{tsd} : \mathcal{O} \to \mathcal{O}_T$ and $\mathrm{ssd} : \mathcal{O} \to \mathcal{O}_S$, that return the task space representation and structure space representation, respectively, of an observation $\omega = \psi(s)$: $\mathrm{tsd}(\omega) = \psi_T(s)$ and $\mathrm{ssd}(\omega) =$

$\psi_S(s)$. We call $\text{tsd}(\omega)$ a *task space descriptor* and $\text{ssd}(\omega)$ a *structure space descriptor*. In a structure space aspectualizable observation space, both functions are aspectualizations.

Example 5.3. Let us assume we have a five-dimensional structure space aspectualizable observation space $\mathcal{O} = \mathbb{R}^5$ with an observation function $\psi : \mathcal{S} \rightarrow \mathbb{R}^5$, $\psi(s) = (x_1, x_2, x_3, x_4, x_5)$. The first three elements of an observation constitute the task space and the latter two the structure space, that is, $\mathcal{O}_T = \mathbb{R}^3$ and $\mathcal{O}_S = \mathbb{R}^2$. $\text{tsd} : \mathbb{R}^5 \rightarrow \mathbb{R}^3$, $\text{tsd}(\omega) = (x_1, x_2, x_3)$ is the task space descriptor and $\text{ssd} : \mathbb{R}^5 \rightarrow \mathbb{R}^2$, $\text{ssd}(\omega) = (x_4, x_5)$ is the structure space descriptor. For an observation $\omega = (5, 2, 1, 8, 0)$ it holds that $\text{tsd}(\omega) = (5, 2, 1)$ and $\text{ssd}(\omega) = (8, 0)$.

5.2.3.2 Relation to the Problem/Agent Space Framework

Concurrently with the author's research on the distinction between goal-directed and generally sensible behavior and the notion of structural similarities among state spaces (Frommberger, 2006) that has been described above, a similar issue was raised by Konidaris (2006). In his framework for transfer in reinforcement learning Konidaris distinguishes between *problem space* and *agent space*. Both contributions have been published at the ICML workshop on "Structural Knowledge Transfer for Machine Learning" in 2006.

In Konidaris's framework, problem space stands for the set of features defining the state space for the overall task, usually the original state space of an MDP. An agent space consists of a set of attributes that are shared under the same semantics by different tasks. The generated feature space is called agent space because it is associated not with the singular tasks, but with the agent itself. Learning takes place in problem space and transfer takes place in agent space.

The idea of a set of features that relates to different tasks and thus is transferable is common the problem/agent space framework and the structure/task space framework presented in this book. Structure space, as defined above, establishes an agent space for structurally similar tasks (called "related tasks" in Konidaris (2006)), but also refers to similarities at different places *within* one task. The proposed structure space aspectualizable observation space is a special case where the agent space representation is a part of the problem space representation. In general, Konidaris (2006) requires no commonalities in problem and agent space.

The distinction between problem and agent space has been used to learn shaping rewards to speed up learning in a new task (Konidaris and Barto, 2006) and to perform knowledge transfer by creating portable options that have been learned in agent space (Konidaris and Barto, 2007).

5.2.3.3 Structure Space Policies

A policy that only operates on $\mathcal{O}_S$ is called a *structure space policy* π_S. It only takes the structure of the state space into account; thus, it implements a generally sensible behavior. The ability to represent structure space in an aspectualizable manner does not automatically lead to a structure space policy, but it gives all the accessibility needed to derive it during or after the learning process, as the next sections show.

A structure space policy, once derived, is applicable in any structurally similar task and enables the agent to operate in a reasonable way. From the agent's observation ω, the structure space description $\mathrm{ssd}(\omega)$ can be achieved by a simple aspectualization, and $\pi_S(\mathrm{ssd}(\omega))$ immediately delivers an action according to a generally sensible behavior.

Thus, a structure space policy is an excellent candidate for transfer. In a new environment, it enables the agent for reasonable operations, which can be the basis for a new policy for the new task or at least lead to proper exploration of the new state space.

In structurally similar structure space aspectualizable state spaces a structure space policy is immediately applicable—no transfer algorithm is needed, as the structure space description is identical over the tasks and can be extracted from the environment without any effort. Thus, there is no loss in the transfer process, because there is no explicit transfer process at all: the structure space policy is simply reused. In learning algorithms, the best transfer is implicit transfer.

5.2.3.4 Domain Knowledge

The existing cross-domain transfer learning algorithms presented in Sect. 3.5.3 all have in common that they rely on predefined relationships between source and target tasks. In other words, the function of how to map from one problem to the other has to be given by an external expert based on *domain knowledge*.

If we have a structure space policy for a state space and want to use it in a structurally similar scenario with a structure space aspectualizable state space, we do not need domain knowledge for transfer: The structure space policy simply applies to the target task. However, domain knowledge is still critical: as pointed out in Sect. 4.5.5, it is essential for building the aspectualizable representation from which we then are able to derive the structure space policy. In other words, the work presented in this thesis uses domain knowledge to achieve expressiveness in the state space representation, and the appropriate representation alone facilitates learning and knowledge transfer.

In the next sections, we investigate two different ways of deriving structure space policies from a learning task. There, we assume that we operate on a structure space aspectualizable observation space $\mathcal{O}$ that is derived from the underlying state space $\mathcal{S}$ with an observation function $\psi(s) = (\psi_T, \psi_S)$. We also assume we have a value function $V : \mathcal{O} \to \mathbb{R}$. Even if Q-learning is used in the practical part of this work, we first regard V instead of Q to facilitate readability. Of course, the considerations are also valid for Q-learning and Q-functions without loss of generality.

5.3 Value-Function-Approximation-Based Task Space Generalization

We first want to investigate how to realize *generalization* in state spaces that are structure space aspectualizable. That is, we want to achieve that the agent is able to reuse knowledge about structurally similar states while learning a single task and thus increase its performance.

5.3.1 Maintaining Structure Space Knowledge

If the learning agent is to be able to reuse previously gathered knowledge when perceiving an observation ω, it should be able to consider the information stored for states that are structurally similar. So first it is necessary to identify structurally similar observations or states. This is straightforward in structure space aspectualizable observation spaces.

Assume we have two states s and s' ($s, s' \in \mathcal{S}$). These states are structurally similar if the observation space $\mathcal{O}$ is structure space aspectualizable and the observations $\omega = \psi(s)$ and $\omega' = \psi(s')$ have an identical structure space representation, that is, $\psi_S(s) = \psi_S(s')$, or, put differently, $\mathrm{ssd}(\omega) = \mathrm{ssd}(\omega')$.

In reinforcement learning, knowledge about states is stored in the value function. Considering all knowledge about structurally similar observations would require looking at all values $V(\omega')$ with $\mathrm{ssd}(\omega) = \mathrm{ssd}(\omega')$ stored in the value function representation. Even in discrete state spaces, this is infeasible, because the value function representation has to be searched in every single step the agent performs. Even with clever access methods, this is too time-consuming. It is more desirable to find a pre-initialized value $V(\omega)$ that can be accessed in constant time even when ω is visited for the first time.

Whenever updating the value function V at an observation ω, we want that not only $V(\omega)$ is updated, but—preferably—*all* $V(\omega')$ with $\mathrm{ssd}(\omega) = \mathrm{ssd}(\omega')$ are updated. If an update of a state also affects all states with the same structure space descriptor (and no states with a different structure space descriptor), we call this *task space generalization*. It can be achieved by using the generalization benefits of value function approximation.

Note that we are operating on a discrete state space, so the values of each observation still can be stored in a table. Here, the focus is on generalization. For this, the value function approximation of *tile coding* is an excellent choice.

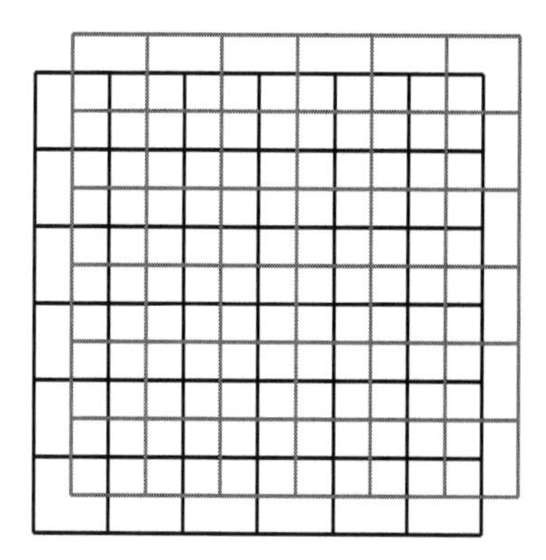

Fig. 5.2 Two overlapping tilings in $\mathbb{R}^2$

5.3.2 *An Introduction to Tile Coding*

This section provides a detailed description of the tile coding mechanism (Sutton, 1996) (which is also known as CMACs) and its application to task space generalization.

5.3.2.1 Tilings

As described in the short introduction in Sect. 3.2.1.4, tile coding partitions the state space into disjoint sets, the receptive fields, or *tiles*, and such a partition is called a tiling. Various partitions are possible: The easiest tiling in 2-D is a regular grid with equally large squares, but also stripes or arbitrary tessellations can form a tiling. Tilings do not need to be symmetrical; they can also be logarithmic, for example.

In a tiling, each $s \in \mathcal{S}$ is assigned to exactly one tile. This is a non-injective function $\kappa : \mathcal{D} \to \mathcal{D}$ and thus a coarsening (see Sect. 4.2.3). In this section, for the sake of simplicity, we only consider the case where each dimension of the state space is partitioned into parts of equal lengths (but those lengths can vary over the partitions). This results in n-dimensional hyper-rectangles.

The resolution of the emerging state space depends on the size of the receptive fields and on the number of tilings. To increase the resolution, several different tilings are used, each with a different offset from the first one, so the tilings are overlapping (see Fig. 5.2). So each state corresponds to one tile within *each* tiling, and due to the offset, and in general, the corresponding tiles differ over the tilings. The larger the tiles, the broader the generalization, and the more tilings we chose, the higher is the resolution of the function approximation.

Neighboring states will more or less map to different tiles over the tilings, depending on their nearness. So information stored for a state s will also affect the stored values for states that are near s—based on the concept of nearness for homogeneous environments in the sense of Roberts (1973) (see Sect. 4.4.1).

5.3.2.2 Storing Values with Tile Coding

A CMAC reduces the state space and provides a discrete state space. Thus, it allows for using a table for value function representation. To lower memory consumption, a hash table is to be preferred.

When using more than one tiling, the values are stored distributed over the tilings. Let n be the number of tilings and $\chi_i : \mathcal{S} \rightarrow \mathcal{S}$ be one of several functions that assign each $s \in \mathcal{S}$ a representative for a tile within a certain tiling i. A hash function $h : \mathcal{S} \rightarrow \mathbb{N}$ now delivers a cell number $h(\chi_i(s))$, and $c : \mathbb{N} \rightarrow \mathbb{R}$ is a function that returns the contents of this cell. A value $V(s)$ equally distributed over the tilings by

$$c(h(\chi_i(s))) = V(s), \quad i = 1, \ldots, n. \tag{5.1}$$

Conversely, a value $V(s)$ is received from the tile coding table as an average over all tilings:

$$V(s) = \frac{1}{n} \sum_{i=1}^{n} c(h(\chi_i(s))). \tag{5.2}$$

The TD-update rule (compare to (2.7)) with a learning rate α and a TD error δ is (in the undiscounted case)

$$c(h(\chi_i(s))) = c(h(\chi_i(s))) + \alpha\delta, \quad i = 1, \ldots, n. \tag{5.3}$$

The following example illustrates how values are stored and how neighboring states are affected:

Example 5.4. To keep the example simple, we assume that our state space is already discrete: let it be a grid world where each cell has an edge length of 1. Cells are aligned with the coordinate axes such that each cell can be represented by the coordinates of one of its corners, $(x, y) \in \mathbb{Z}^2$. For this state space, we choose two tilings representing V_1 and V_2 with a tile size of 2×2 cells each, that is, each tile covers four cells. The tilings are shifted with an offset of 1 in each dimension. The value function V is derived from V_1 and V_2 as described in (5.2). To keep the example simple, we ignore the hash function here. We also assume a learning rate of $\alpha = 1$. In the beginning, V_1 and V_2 are initialized with 0, so that $V = \frac{1}{2}(0+0) = 0$ for all (x, y):

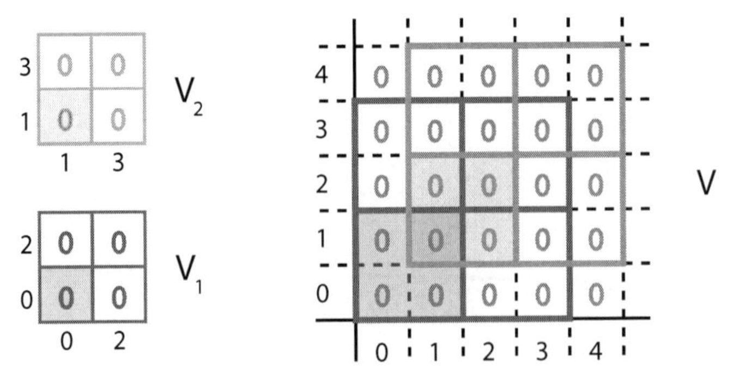

In a first step, we want to update $V(1,1)$ to a value $V'(1,1) = 4$. This results in a TD-error $\delta = V'(1,1) - V(1,1) = 4 - 0 = 4$. The coordinate $(1,1)$ refers to $V_1(0,0)$ and $V_2(1,1)$. The affected cells are marked in the tables above. V_1 and V_2 are updated according to (5.3): $V_1(0,0) := V_1(0,0) + \alpha\delta = 0 + 1 \times 4 = 4$, $V_2(1,1) := 0 + 4 = 4$. We now get the following allocation of V_1 and V_2 from which the corresponding values for V can be computed:

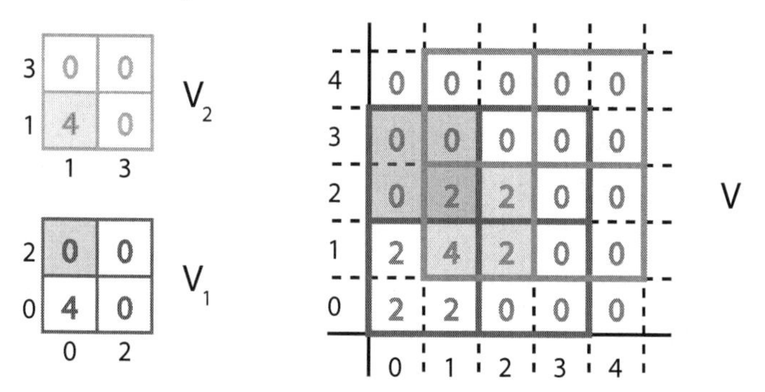

For example, $V(1,2) = \frac{1}{2}(V_1(2,2) + V_2(0,0)) = 2$.

As a next update we observe $V'(1,2) = 4$. Now, $\delta = V'(1,2) - V(1,2) = 2$, so $V_1(2,2) = 0 + 2 = 2$ and $V_2(0,0) = 4 + 2 = 6$. The new allocation of V_1, V_2, and V is now as follows:

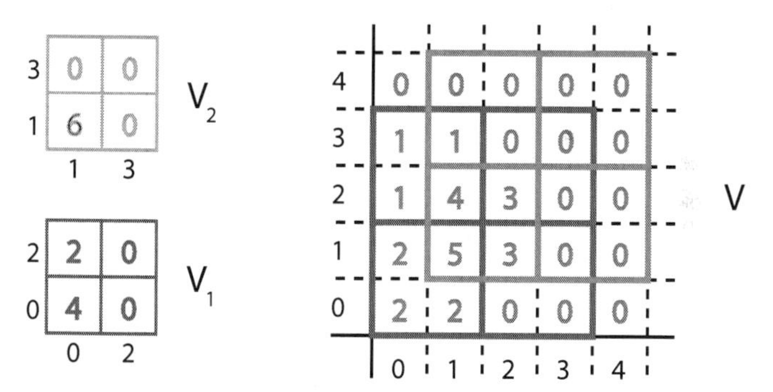

Now $V(1,1) = 5$ and $V(1,2) = 4$. We also encounter, for example, $V(2,1) = 3$, even if $(2,1)$ has not been visited at all.

5.3.2.3 Tile Coding and Convergence

There is no guarantee of convergence for Q-learning with tile coding (Baird, 1995). Gordon (1995) was able to show that CMACs can diverge in special cases and uses the term *exaggerator* for it, because the TD-error can in certain cases be increased instead of being lowered, which may lead to a less robust learning behavior.

A variant of tile coding, the so-called *averager CMAC* or *A-CMAC* (Timmer and Riedmiller, 2007), is a more local procedure that addresses the exaggerator problem.

It calculates a tile-specific TD-error δ_i for every singular tile, so that the update rule (5.3) changes to

$$c(h(\chi_i(s))) = c(h(\chi_i(s))) + \alpha\delta_i, \quad i = 1,\ldots,n. \tag{5.4}$$

This leads to an increase in stability over the learning process, but with some loss in generalization properties. However, this can even be desirable in certain cases.

Q-learning on A-CMACs has been shown to converge to an optimal solution under certain circumstances (Timmer and Riedmiller, 2007).

5.3.2.4 Runtime and Memory Considerations

Tile coding tables can be appropriately stored in a hash table, which limits the amount of memory allocated to what is really needed to store the value function. However, the size of the hash table has to be given in advance and has to be chosen large enough to avoid an excessive number of hashing collisions. Given an appropriately designed hash function, accessing a value has an amortized access time of $O(1)$ on average.

For determining the corresponding tile $h(\chi_i(s))$ for a state s we assume a runtime of $O(1)$. Aside from very exotic tilings, χ_i is an arithmetic function with constant runtime, typically some sort of rounding such as $\chi_i(s) = \lfloor \frac{s}{c} \rfloor$ with a constant $c \in \mathbb{R}$, which assigns each s to a hypercube with an edge length of c in constant time.[1]

For both reading and storing a value in the tile coding table, one hash table access is needed for each of the n tilings, so both read and write access need $O(n)$ time. Updates require one read and one write access; thus, the runtime complexity of updates is $O(n)$ as well. Because the number of tilings n is constant, the requirement of constant time access is met.

The same holds for memory consumption. We can regard each tiling as a table of its own, so the memory effort for function approximation with tile coding with n tilings is also $O(n)$. Thus, this method is suitable for inclusion in reinforcement learning algorithms.

5.3.3 *Task Space Tile Coding*

The following section introduces a method to use tile coding in order to achieve task space generalization within tasks as proposed in Sect. 5.3.1. We refer to this procedure as *task space tile coding* (TSTC).[2]

[1] We assume here that the floor function $f(x) = \lfloor x \rfloor$ can be computed in $O(1)$.

[2] The effect of task space tile coding was first described under the term GRLPR in some earlier publications (Frommberger, 2006, 2007b, 2008a). There, it was interpreted as a special variant of the RLPR representation (which will be introduced in Section 6.3.1) more than as a general method in structure space aspectualizable state spaces.

5.3.3.1 Design of the Tiling Function

First, we make some assumptions on the design of the observation space. All these assumptions can easily be fulfilled by an appropriate application of abstraction paradigms. We have a structure space aspectualizable observation space $\mathcal{O} = \mathcal{O}_T \times \mathcal{O}_S$. Furthermore, we assume both task space and structure space to be finite and discrete—a property that we desire for state space representations anyway. However, the proposed method can also be applied to continuous state spaces. W.l.o.g. we assume that $\mathcal{O}_T \subset \mathbb{N}^k$ and that each $p \in \mathcal{O}_T$ has the form $p = (p_1, \ldots, p_k)$. Furthermore, we assume w.l.o.g. that consecutive numbers are assigned observed features $p_i \in \mathbb{N}$. For every dimension i in $\mathcal{O}_T$ there exists a $p_i^{\max}$ with $p_i^{\max} \geq p_i$, that is, every feature within p has a maximum value. The same holds for a minimum value. For the sake of simplicity of notation this minimum value is assumed to be 1 for each dimension. A vector $p^{\max} = (p_1^{\max}, \ldots, p_n^{\max})$ gathers the maxima for each dimension.

For structure space $\mathcal{O}_S$ it is also assumed w.l.o.g. that $\mathcal{O}_S \subset \mathbb{N}^m$. Each $q \in \mathcal{O}_S$ has the form $q = (q_1, \ldots, q_m)$. For each dimension there is a minimal distance between the possible features. From that emerges a distance vector $d = (d_1, \ldots, d_m)$ with

$$d_i = \min_{q, \overline{q} \in \mathcal{O}_S, q \neq \overline{q}} (|q_i - \overline{q}_i|). \tag{5.5}$$

Let us now define a tiling function $\chi : \mathcal{O} \to \mathcal{O}$ as follows

$$\begin{aligned} \chi(\omega) &= \chi(p_1, \ldots, p_k, q_1, \ldots, q_m) && (5.6)\\ &= (p'_1, \ldots, p'_k, q'_1, \ldots, q'_m) && (5.7)\\ &= \left(\frac{p_1}{p_1^{\max}}, \ldots, \frac{p_k}{p_k^{\max}}, \frac{q_1}{d_1}, \ldots, \frac{q_m}{d_m} \right). && (5.8) \end{aligned}$$

Obviously, it holds that $p'_i = \frac{p_i}{p_i^{\max}} \in (0, 1]$. Furthermore, for two different structure space representations q_i and q_j $(i \neq j)$ it holds that $|q'_i - q'_j| = |\frac{q_i}{d_i} - \frac{q_j}{d_j}| \geq 1$, that is, the distance between different features in structure space is always at least 1. So in a hypercube with an edge length of 1 in each dimension, all instances of $(p'_1, \ldots, p'_k, q'_1, \ldots, q'_m)$ map to a single hypercube cell, while no two different q_i and q_j are mapped to the same interval in any dimension. Thus, different structure space descriptors q are always mapped to different cells.

Even if it occurs that $p_i^{\max}$ cannot be distinguished in advance, this does not break task space tile coding. In this case, generalization will only apply to a subset instead of to *all* structurally similar states. However, this is not a severe limitation, because non-local generalization is still achieved within the subsets.

5.3.3.2 Maintaining the Original Resolution

Mapping states to tiles reduces the granularity of the observation space within each tiling. However, we do not want to experience a further loss of resolution in the observation space which is assumed to be chosen appropriately for the problem at hand. To ensure having a sufficiently large resolution to store individual values $V(\omega)$ for any $\omega \in \mathcal{O}$ we need to choose enough tilings. A number of $n = \max_i p_i^{\max}$ tilings can provide this requirement and conserves the initial resolution of the discrete observation space, so that each $\omega \in \mathcal{O}$ can still be distinguished between within the value function representation.

5.3.3.3 Achievements

The following insights can be obtained from the consideration about how values are stored within the tile coding table:

- Generalization is exclusively applied to task space $\mathcal{O}_T$; structure space $\mathcal{O}_S$ is not matter of generalization.
- An update of a value affects *all* states with the same structure space representation.
- Updates affect every observation, even those that have not been perceived and updated before.
- All elements of the observation space $\mathcal{O}$ can still be distinguished between and have an individual value assigned.

As a consequence, the presented method fulfills the requirements for non-local in-task generalization. Whenever the agent encounters a new observation with an already known structure space descriptor, it will be able to refer to already gathered knowledge. The generalization abilities of this method can be directly utilized within the running task.

An update for one observation affects all structurally similar observations and enables generalization, but the generalization is not smooth, that is, states with the same distance from the original observation may encounter different amounts of generalization. This is due to two reasons. First, the amount of generalization depends on the similarity of the structure space representation. This is a welcome feature under the assumption that similar task space representations require the same action with a high probability. If that is not the case, the amount of generalization is arbitrary, so generalization may work better in similar task space representations than in dissimilar ones. Of course, this kind of generalization is valuable nonetheless, as differences will be smoothed over the learning process.

The second reason emerges from the tessellation structure of the tilings. Rewards are not distributed continuously over the dimensions. This is a limitation caused by the tile coding mechanism. But this shortcoming is acceptable as well, because occurring differences will also be wiped away while learning proceeds and more different observations are made that affect different tiles. The simplicity of the tile coding mechanism compensates for that.

5.3.4 Ad Hoc Transfer of Policies Learned with Task Space Tile Coding

The last section described the effect of task space tile coding within the current task: Due to the explicit representation of structure space, generalized knowledge is made available for rarely or never visited states. The same holds for unknown states in another, but structurally similar environment: Whether the newly perceived observation is from the original task or from the new task does not matter for an appropriate action selection. So if both environments share a structure space, the resulting policy from the source task implements a generally sensible behavior also in the target task, because the generalized knowledge applies to all states, including unperceived ones. Thus, no effort has to be taken for transferring knowledge from source to target task, the learned policy can be immediately applied. Therefore, we refer to transfer of policies learned with function approximation-based task space generalization as *ad hoc transfer*.

Two prerequisites have to be ensured for successful ad hoc transfer:

1. No two task space representations in source and target space may be identical. That is, given source and target tasks with observation spaces $\mathcal{O}$ and $\mathcal{O}'$ and the corresponding task spaces $\mathcal{O}_T$ and $\mathcal{O}'_T$, it must hold that $x \in \mathcal{O}_T \Leftrightarrow x \notin \mathcal{O}'_T$. This is important, because within a new environment task space information is not relevant anymore, and a state with the same task space descriptor will refer to values stored for an unrelated task. It must be ensured that all task space descriptors are newly created. For example, this can be achieved by assigning a fixed offset when enumerating features.
2. To make sure that generalization also affects the task space of the target task, $p^{\max}$ has to be chosen appropriately. This has to be considered *before* starting to solve the source task. Each $p_i^{\max}$ has to be chosen greater than the maximum observable feature in both source and target tasks. As features in both tasks have to be represented mutually exclusively (see above), it must hold that $p_i^{\max} \geq \max(p_i, p'_i)$ with $p \in \mathcal{O}_T$ and p' in $\mathcal{O}'_T$. In discrete observation spaces, choosing a large value for $p^{\max}$ is not critical, because each discrete feature is represented on its own when a sufficient number of tilings is ensured (see Sect. 5.3.3).

Both prerequisites have to be considered before learning the source task. However, ensuring these properties is easy and does not reveal additional problems.

When the prerequisites are met, task space information of the target task can be included in the originally learned Q-function such that task space knowledge from the source task is maintained. This can be seen as a contribution to the concept of *lifelong learning*, as new knowledge is added while the structure space knowledge is continuously refined over every newly learned task.

5.3.5 Discussion of Task Space Tile Coding

Task space tile coding is a very efficient means to provide generalizing behavior over structural properties within a task while maintaining the original resolution of a discrete state space. Thus, it allows for increased learning performance. Furthermore, ad hoc transfer is a stunning means to reuse knowledge in another task, because the transfer process reduces to just taking and using a previously learned policy. No computational effort is needed for this kind of transfer learning at all, the desired knowledge is immediately present.

While it is universally applicable for knowledge transfer, two drawbacks can be identified in the ad hoc concept of structure space transfer, which are caused by the way generalization is achieved:

1. First, in the target environment or at unknown places in the source environment, the system is confronted with completely unknown task space representations and usually returns a reasonable reward. However, the reward is not independent of the newly perceived task space representation, and different inputs may result in different actions, even if the structure space representation is the same.
2. Second, the system has to be trained with discounted rewards (see Sect. 2.3.2), or with negative reinforcement for any action taken. That means that Q-values are high for observations near the goal state (and any other state providing a positive or negative reinforcement) and decrease with distance from it. As tile coding sums up values over different tilings, places in the vicinity of the goal have the biggest impact on the generalized strategy. This phenomenon is to be expected with any function approximation. If the resulting structure space policy is influenced by properties of task space in this way, we call this a *task-space-biased generalization*.

Overall, ad hoc knowledge transfer with task space tile coding will deliver a good base for a further learning task, but it may be not directly applicable, because the named drawbacks can disturb the derivation of a generally sensible behavior in a way that the transferred strategy may not prove to be immediately successful all the time. So task space tile coding is best suited for in-task generalization (see Sect. 7.2.4 for an empirical evaluation) and as a basis for further learning in a target task. Certainly, ad hoc transfer is stunningly simple and worth consideration. As it does not need any computational time, its effect can be assessed very easily.

For deriving an universally working policy in the target task we take a look at further possibilities in the next sections.

5.4 A Posteriori Structure Space Transfer

The following section presents an algorithm for structure space policy transfer that aims at avoiding a task space bias. It is an a posteriori method that operates on a successfully learned value function and policy. Thus, it is applicable only to transfer, and not to in-task generalization. We refer to this procedure as *a posteriori structure space transfer* (APSST).[3]

In this section, we assume that the source task is learned with Q-learning, and so regard a Q-function $Q : \mathcal{O} \times \mathcal{A} \to \mathbb{R}$ instead of a value function here. Again, the prerequisite for this algorithm is the existence of a structure space aspectualizable observation space.

5.4.1 Q-Value Averaging over Task Space

We have a structure space aspectualizable observation space $\mathcal{O} = \mathcal{O}_T \times \mathcal{O}_S$, a POMDP $\langle \mathcal{O}, \mathcal{A}, T, R \rangle$ and a policy π operating on a Q-function $Q : \mathcal{O} \to \mathbb{R}$. We now want to achieve a structure space policy π_S with a function $Q_S : \mathcal{O}_S \to \mathbb{R}$ operating on the structure space of the source task. Each target task can then use π_S.

To generate a structure space policy, information on observations with the same structure space descriptor is summarized. This is achieved by *Q-value averaging*: To get the new Q-function $Q_S(\overline{\omega}, a)$ ($\overline{\omega} \in \mathcal{O}_S$, $a \in \mathcal{A}$), we calculate the arithmetic means over all the Q-values of all observation-state pairs $(\omega, a) = ((\mathrm{tsd}(\omega), \overline{\omega}), a)$ ($\omega \in \mathcal{O}$) with the same structure space representation $\overline{\omega} = \mathrm{ssd}(\omega)$:

$$Q_S(\overline{\omega}, a) = \frac{1}{|\{x \in \mathcal{O}_T | Q((x, \overline{\omega}), a) \neq 0\}|} \sum_{y \in \mathcal{O}_T} Q((y, \overline{\omega}), a). \tag{5.9}$$

We only regard observations we have information on, that is, where the Q-value does not equal 0.

However, iterating over the whole task space may be computationally infeasible, even in discrete observation spaces. For example, in a seven-dimensional task space with 20 possible instances of features in each dimension and an action space size of 3 (these numbers refer to a learning task described in Sect. 7.4.1), 3.8×10^{10} additions would be needed to compute Q_S. But usually only a small fraction of the theoretical task space is physically existent. Furthermore, it may not have been completely explored. So in practice it is only necessary to keep track of the visited state-action pairs and to regard those. In all the examples shown later in this work, the algorithm terminated within at most a few seconds on usual desktop computers.

5.4.2 Avoiding Task Space Bias

The algorithm as presented in the previous section is still task-space-biased because the problem of discounted rewards is not being taken into account yet. Therefore, we extend (5.9) by a weighting factor to even out the differences in Q-values for

[3] This section presents a significantly revised version of preliminary work on APSST that has been published in Frommberger (2007a).

states at different distances from the goal state. We can compare the Q-values for all the actions a at an observation ω and normalize them with regard to the absolute value of the maximum Q-value $|\max_{b \in \mathcal{A}} Q(\omega, b)|$. Then for the normalized values the following holds:

$$-1 \leq \frac{Q(\omega, a)}{|\max_{b \in \mathcal{A}} Q(\omega, b)|} \leq 1 \quad \forall a \in \mathcal{A}, \omega \in \mathcal{O}. \tag{5.10}$$

Thus, the overall equation to generalize to a structure space policy is

$$Q_S(\overline{\omega}, a) = \frac{1}{|\{x \in \mathcal{O}_T | Q((x, \overline{\omega}), a) \neq 0\}|} \sum_{y \in \mathcal{O}_T} \frac{Q((y, \overline{\omega}), a)}{\max_{b \in \mathcal{A}} |Q((y, \overline{\omega}), b)|} \tag{5.11}$$

for every $\overline{\omega} \in \mathcal{O}_S$.

An algorithmic notation of the procedure is given in Algorithm 2. It iterates over the set $\mathcal{O}_v$ of all visited observations and for each them over all actions, so its runtime is $O(|\mathcal{O}_v| \times |\mathcal{A}|)$. The final loop iterates over the newly generated set of visited structure space representations, which is smaller than $\mathcal{O}_v$, so the linear effort is not affected. As $|\mathcal{A}|$ is constant, the a posteriori structure space generalization algorithm has a linear runtime with regard to the number of visited observations.

Algorithm 2 A posteriori structure space generalization

Require: $\mathcal{O}_v$ contains all visited observations
Require: T is an empty associative table
  $M_S \leftarrow \emptyset$
  for all $\omega \in \mathcal{O}_v$ **do**
    $i \leftarrow 0$
    for all $a \in \mathcal{A}$ **do** ▷ look at the Q-values for each action
      $v[i] \leftarrow Q(\omega, a)$
      $i \leftarrow i + 1$
    end for
    $m \leftarrow \max_j |v[j]|$ ▷ maximum for weighting factor
    for all $a \in \mathcal{A}$ **do**
      $Q_S(\text{ssd}(\omega), a) \leftarrow Q_S(\text{ssd}(\omega), a) + \frac{1}{m} Q(\omega, a)$
    end for
    $M_S \leftarrow M_S \cup \text{ssd}(\omega)$ ▷ store processed structure
    $T[\text{ssd}(\omega)] \leftarrow T[\text{ssd}(\omega)] + 1$ ▷ count appearance
  end for
  for all $\overline{\omega} \in M_S$ **do** ▷ average Q-values
    $Q_S(\overline{\omega}) \leftarrow Q_S(\overline{\omega}) / T[\overline{\omega}]$
  end for

5.4.3 Measuring Confidence of Generalized Policies

When generating a structure space policy the aim is that the resulting strategy assigns to any observation the action that fits best *in general* and yields a generally sensible behavior. For many structure space representations, there will be one action that is most sensible independent of task space, but this may not necessarily be the case. In the office environment robot navigation task, for example, the actions to take at a right turn are clear, but at a crossing, making each turn or heading straight might be equally good options.

Also, an intermediate policy from a learning process that has not yet finished will represent incomplete knowledge and thus be inconsistent over different observations with the same structure space representation. This kind of inconsistency, caused by an unfinished learning process, will be leveled out while learning proceeds. In other words, the agent will become more *certain* of which actions to take.

Applying and evaluating different policies that are candidates for knowledge transfer in the target domain can be an expensive and time-consuming procedure. Thus, an a priori evaluation of the quality of the policies is helpful. The certainty in action selection is an important hint for the expected quality, because one should be interested in achieving a structure space policy from a strategy the agent is certain of. We can measure this by taking a look at the Q-values for different actions at a given observation. The greater the difference between the highest and the remaining Q-values, the higher is the confidence of which action to take at this observation.

So for an $\overline{\omega} \in \mathcal{O}_S$, $\overline{\omega} = \text{ssd}(\omega)$ we define a *decision confidence* $\text{conf} : \mathcal{O}_S \to \mathbb{R}$:

$$\text{conf}(\overline{\omega}) = \frac{1}{|\mathcal{A}|} \sum_{a \in \mathcal{A}} (\max_{b \in \mathcal{A}}(Q_S(\overline{\omega}, b)) - Q_S(\overline{\omega}, a)). \tag{5.12}$$

If there is no information collected for an action a yet, the result is too arbitrary. In this case, the decision confidence is set to 0.

Because of the mentioned situations where two or more actions are equally appropriate, this measure particularly makes sense for evaluation when summed up over the whole observation space for the whole strategy. So we define the decision confidence for a structure space policy π_S established by Q_S on a structure space $\mathcal{O}_S$ as the average over all decision confidences:

$$\text{Conf}(\pi_S) = \frac{\sum_{\overline{\omega} \in \mathcal{O}_S} \text{conf}(\overline{\omega})}{|\{\overline{\omega} | \text{conf}(\overline{\omega}) \neq 0\}|}. \tag{5.13}$$

Only observations with a non-zero decision confidence are considered ($\text{conf}(\overline{\omega}) \neq 0$), that is, only those of which we already have sufficient knowledge.

In the case of successful learning, $\text{Conf}(\pi_S)$ will increase over the learning process and converge to a certain value. It can then be used to compare and evaluate different structure space policies before applying them to the target space: Policies with higher decision confidence can be expected to be applicable more successfully because only they do not show insecurity due to insufficient learning, and only insecurity with regard to the environment's structure remains.

Decision confidence is no absolute measure. It can only be compared for policies learned in the same MDP, because structural properties of the state space and the design of rewards have a significant influence on it.

Decision confidence can easily be computed while computing structure space policy as presented in Algorithm 2, so it can be acquired as a side product of this procedure. However, it can also be computed on its own. The resulting algorithm works exactly as Algorithm 2 and thus shares the runtime considered in Sect. 5.4.2.

5.5 Discussion of the Transfer Methods

Both task space tile coding (TSTC, Sect. 5.3) and a posteriori structure space transfer (APSST, Sect. 5.4) are able to create a structure space policy that is applicable in structurally similar tasks. This section compares the methods with regard to a possible application and takes a look at possible further developments.

5.5.1 Comparison of the Transfer Methods

TSTC generates a structure space policy while learning the original task, so it is immediately applicable without additional computational effort. In TSTC the overall policy works as a structure space policy when confronted with an unknown task space descriptor. The advantage of TSTC is that the original Q-function is maintained and new task spaces can be learned by updating one and the same Q-function. Furthermore, TSTC provides in-task generalization, but is task-space-biased.

Table 5.1: Comparison of the generalization methods with regard to certain properties: "+" stands for provided, "−" for not provided, and "∘"' for partially provided

	TSTC	APSST
optimal solution	−	+
in-task generalization	+	−
no task space bias	−	∘
immediately applicable	+	−
pure structure space policy	−	+
original knowledge maintained	+	−
no additional parameters	+	+
confidence measures	−	+

APSST avoids task space bias as far as possible (see Sect. 5.4.1) but it requires extra, yet low computational effort to create the strategies, and no in-task generalization is achieved. Certainly, this is also beneficial, because the originally learned policy has been learned as it is and not been affected by function approximation in TSTC, which can be expected to lead to a slight loss in the quality of the overall solution. APSST is furthermore able to offer generalization assessment by providing confidence measures on the structure space policy.

Summed up, when looking for in-task generalization and rapid learning progress, TSTC is the method of choice. To achieve a "clean" structure space policy, APSST is the most appropriate. Table 5.1 gives a schematic overview of the properties named in this section. An empirical comparison of transfer capabilities of TSTC and APSST in a robot navigation scenario is given in Sect. 7.4.2.

5.5.2 Outlook: Hierarchical Learning of Task and Structure Space Policies

If task and structure space can be identified and represented separately, it is possible to regard task space and structure space as two different learning problems. Both the problems are not Markov, and, in general, neither of their solutions solves the overall task. However, we could learn two separate policies, a task space and a structure space policy, at the same time to achieve two separate Q-functions, Q_T and Q_S, for task space and structure space.

At each time step we now have access to both Q-functions which represent the best choice of action regarding both task space and structure space. These actions, of course, can differ, so we need a mechanism to decide how to choose an action based on the Q-values of task and structure space policies. Accordingly, rewards have to be assigned to both Q_T and Q_S. Thus, a decision module has to access both task and structure space Q-functions such that we get a hierarchical architecture. The decision module fulfills two purposes: First, it computes an overall Q-value from Q_T and Q_S, and second, it distributes the reward to both Q-functions.

A straightforward way to implement such a decision module would be to have a weighting parameter $\mu \in [0,1]$ to weight between the importance of task and structure space, so that $Q(s,a)$ is calculated as follows

$$Q(s,a) = \mu Q_T(s,a) + (1-\mu)Q_S(s,a) \tag{5.14}$$

In the same manner, rewards could be split for both task and structure space Q-functions with R_T being the reward to update Q_T and R_S being the reward to update Q_S, respectively, according to the Q-learning update rule given in (2.11):

$$R_T(s,a) = \mu R(s,a), \tag{5.15}$$

$$R_S(s,a) = (1-\mu)R(s,a). \tag{5.16}$$

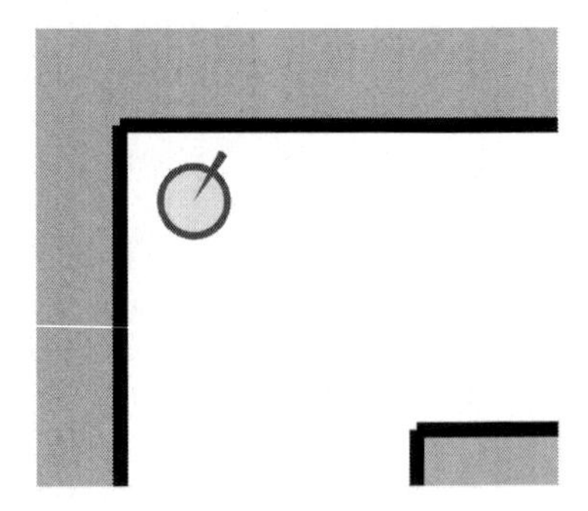

Fig. 5.3 A robot in a corner of a corridor: A right turn is the only reasonable action here to prevent a collision

Given that, we achieve a structure space policy at no price, as Q_S can directly be taken and applied to a structurally similar task.

In this architecture, the choice of μ is the crucial part. If it is chosen too high, the system performs badly with respect to task space generalization; this will result in unnecessarily long learning times. If it is chosen too low, the structure space might be too prominent and override task space decisions at important states, so a successful solving of the task cannot be achieved.

In first experiments, this hierarchical approach proved to be successful in learning scenarios as used in Chap. 7. Learned structure space policies also applied well to new environments. However, a thorough investigation of the properties of such a hierarchical method will be necessary.

5.6 Structure-Induced Task Space Aspectualization

Apart from the application of structure space policies in new and unknown tasks and in-task generalization properties, the knowledge about generally sensible behavior in structure space aspectualizable state spaces can also be beneficial for manipulating the observations of the agent in order to achieve a more manageable state space.

A central question of abstraction is that of which information to retain and which to drop—and not everything is important under *any* circumstances. Depending on the situation, different levels of abstraction can be appropriate. While a detail may be critical for action selection in one situation, it can be completely irrelevant in another. In fact, it can be the case that the structure of the state space alone enforces a particular action. For example, when a robot navigates too closely into a corner such that any action besides a right turn would lead to a collision with a wall, this right turn is the only reasonable action (Fig. 5.3). This action is enforced by the world's structure, which does not allow for a change of decision before the robot reaches a more open space. Before that *decision point* the knowledge of the concrete whereabouts of the turn within the world does not matter for action selection—actions are *only* induced by the world's structure at this point in time.

5.6.1 Decision and Non-decision States

In many tasks, there are situations where the choice of actions only depends on structural information. In other words, when structure space strictly enforces a certain action selection, task space information becomes irrelevant for the control process. These *non-decision states* have a *non-decision structure*.

That idea leads to the *structure-induced task space aspectualization* (SITSA) algorithm.[4] We can shrink the state space if we simply *ignore* task space information at any non-decision state—such that the action selection there is only based on the structural representation.

5.6.2 Identifying Non-decision Structures

SITSA requires knowledge about which $\omega \in \mathcal{O}$ are non-decision states, so it is necessary to identify non-decision structures. Unfortunately, we usually cannot identify non-decision structures in advance. To detect them, one can resort to knowledge gained in a previous task (the source task) that shares the same structure space. If $\omega_0 \in \mathcal{O}$ is a non-decision state, then all $\omega \in \mathcal{O}$ with $\mathrm{ssd}(\omega) = \mathrm{ssd}(\omega_0)$ are also non-decision states. Thus, non-decision states are identified by their structure space representation.

In any non-decision state, the agent is certain which action to take, as there is only one reasonable choice that leads to a high reward expectation. That means that we encounter the Q-value for one action as being significantly higher than that for any other action. In the example of the robot in the corner given above, the Q-values for anything but turning right would be rather low because the action would result in a collision, but the Q-value for turning right would be quite high because this action can enable the robot to reach its goal in the end.

To identify such non-decision structures, we take a structure space policy Q_S learned in a source task derived with Algorithm 2 (see Sect. 5.4). We now again iterate over structure space and look at the normalized averaged values of Q_S to determine the difference between the highest and second highest values of Q_S for different actions, that is, to check whether one action leads to outstandingly high Q-values in the original Q-function. This defines a *decision certainty function* $\mathrm{cert} : \mathcal{O}_S \to \mathbb{R}$:

$$\mathrm{cert}(\overline{\omega}) = \frac{1}{|\mathcal{A}|}(\max_{b \in \mathcal{A}} Q_S(\overline{\omega}, b) - \operatorname*{smax}_{b \in \mathcal{A}} Q_S(\overline{\omega}, b)) \tag{5.17}$$

with smax denoting a function for “second highest value” to facilitate readability. This is similar to the confidence measure $\mathrm{conf}(\overline{\omega})$ defined in Sect. 5.4.3.

---

[4] The SITSA algorithm has been published in Frommberger (2009). Some ideas of SITSA already appeared in a landmark selection algorithm published in Frommberger (2008b).

A very high value of $\mathrm{cert}(\overline{\omega})$ denotes that one specific action was definitely preferred at states with a certain structural representation $\overline{\omega}$ and thus is an indicator for $\overline{\omega}$ being a non-decision structure.

5.6.3 SITSA: Abstraction in Non-decision States

Using (5.17) we first identify a set $\mathcal{O}_{\mathrm{NDesc}} \subset \mathcal{O}_S$ of obvious non-decision structures. It contains all structures where $\mathrm{conf}(\mathrm{ssd}(s)) > c_{\min}$, with $c_{\min} \in \mathbb{R}$ being a predefined constant that defines the minimal needed decision certainty that qualifies for a non-decision structure.

We abstract from task space information at non-decision states, because it is not needed there. This is achieved by enforcing the same unique task space representation at any state with a non-decision structure such that $\mathrm{tsd}(\omega) = \perp$ for all $\omega \in \mathcal{O}$ with $\mathrm{ssd}(\omega) \in \mathcal{O}_{\mathrm{NDesc}}$ (with $\perp$ denoting a predefined "empty" representation). All other states remain unmodified. Formally, we obtain an abstraction function $\Theta : \mathcal{O}_T \times \mathcal{O}_S \to \mathcal{O}_T \cup \{\perp\} \times \mathcal{O}_S$:

$$\Theta(\omega) = \begin{cases} (\perp, \mathrm{ssd}(\omega)) & \text{if } \mathrm{ssd}(\omega) \in \mathcal{O}_{\mathrm{NDesc}} \\ o & \text{else} \end{cases} . \tag{5.18}$$

Θ abstracts from task space features depending on the situation given by knowledge about the structure of the environment and thus shrinks state space size. At any non-decision state now only structural information is represented and anything else is omitted. Thus, we encounter a generalization effect, because structurally identical locations (that is, states ω and ω' with $\mathrm{ssd}(\omega) = \mathrm{ssd}(\omega')$) now share the same abstract state $(\perp, \mathrm{ssd}(\omega))$, and thus learned knowledge applies to various places within $\mathcal{O}$. This method is called *structure-induced task space aspectualization* (SITSA).

5.6.4 Discussion of SITSA

Generalization certainty can only be calculated over an existing policy, as it makes use of the differences of Q-values at structurally similar states. Even if structure space policies and generalization certainty can easily be computed during the learning task, SITSA is not meant to be applied after the learning process starts. The reason is that SITSA changes the state space, and thus the state space would then be expanded instead of shrunk in this case.

However, there are exceptions where it makes sense to first start a learning run on the original observation space $\mathcal{O}$, calculate a generalization confidence after some time, and then continue under SITSA. This could be an option in very large state spaces where a good structure space policy can be achieved very early before the overall task can be solved, so that it can be beneficial to start the task on $\mathcal{S}$ and then continue on $\Theta(\mathcal{S})$. The crucial part, of course, is to reliably identify those exceptional tasks.

A prerequisite for identifying non-decision structures as described in Sect. 5.6.2 is that the consequences of the actions $a \in \mathcal{A}$ that are used in the source task be considerably different. If we had finer-grained actions, non-decision structures might not be identified successfully. Thus, it is reasonable to restrict to few, but different actions in the source task. In the task where SITSA is applied, a finer granularity of actions can then be made available, and different actions can be favored for non-decision states than those learned in the source task.

SITSA, as presented in this section, needs a discrete state space, because it iterates over different state space representations. Its main purpose is to add to an existing abstract discrete representation. But SITSA can be adapted to work with continuous state spaces as well if a value function approximation is used that allows us to assign a representative with the same internal representation to any observed state. For example, this is the case for value function approximation with CMACs (see Sect. 5.3.2).

An application of SITSA in a robot control context called SDALS is presented in Sect. 6.2.3.3 and evaluated in Sect. 7.2.7.

5.7 Summary

Reuse of knowledge is a central concept to allow for learning complex control tasks with reinforcement learning. Generalization enables an agent to benefit from gained knowledge at different locations within the state space while learning, and transfer learning relates to the question of how to utilize this knowledge to learn tasks in new environments more efficiently. In this chapter we investigated the properties of observation space representations with regard to generalization and transfer learning capabilities. It has been worked out that structural similarity is the main concept that connects two domains or different parts of the state space within the same task.

In sequential decision problems, we can distinguish between goal-directed and generally sensible behavior, which corresponds to the distinction between task space and structure space. Generally sensible behavior operates on structure space, and it is this aspect of knowledge that makes an excellent candidate for transfer. So-called structure space policies are applicable to any structurally similar environments.

Domain knowledge is used to achieve expressiveness in the state space representation, and the appropriate representation alone facilitates learning and knowledge transfer. When choosing representations that are aspectualizable with regard to structure space, both task and structure space descriptions can be accessed easily. The concept of structure space aspectualizable state spaces was introduced: Structure space aspectualizable state space representations are the key to successful reuse of knowledge in and across tasks.

Two different transfer methods have been derived. Task space tile coding (TSTC) enables for non-local in-task generalization and ad hoc transfer to new environments. A posteriori structure space transfer (APSST) can only be applied for new tasks, but is not affected by task space bias. It also allows for deriving confidence values to denote how sure the structure space policy is about decisions.

Structure space aspectualizable state spaces allow for selective consideration of task space knowledge based on knowledge derived from previous learning tasks. This also enables structure-induced task space aspectualization (SITSA), a method that uses structure space policies generated with APSST to find decision structures within the structure of the environment.

Chapter 6
RLPR – An Aspectualizable State Space Representation

In this chapter, we get to know a qualitative spatial representation tailored for indoor robot navigation tasks. First, the task to be solved is introduced and task space and structure space are identified along with considerations about frames of reference (Sect. 6.1). In Sect. 6.2 we derive a suitable representation for task space based on the relative positions of landmarks. Selection of landmarks is also discussed there, and two alternatives for landmark consideration (point-based landmarks and sampled views) are introduced. For structure space, the relative line position representation RLPR is introduced in Sect. 6.3. Section 6.4 brings together task and structure space representations to form landmark-enriched RLPR. After a discussion of robustness of this representation in Sect. 6.5, the chapter closes with a summary.

6.1 Building a Task-Specific Spatial Representation

Based on the insights on abstraction and knowledge transfer gained in the preceding two chapters, we now derive a qualitative abstract state space representation tailored for a particular control task. As argued before, the representation has to be structure space aspectualizable.

As pointed out by Sloman (1985), specialized formalisms for different purposes are to be preferred to those trying to cover all knowledge in a single representation. Expert knowledge on the task to solve helps us find a suitable description of the problem.

The task regarded here is a goal-directed indoor robot navigation task. Although this task is associated with a continuous state space that can become very large, this problem is often tackled by reinforcement learning approaches. It is introduced in the following.

L. Frommberger, *Qualitative Spatial Abstraction in Reinforcement Learning*, Cognitive Technologies, DOI 10.1007/978-3-642-16590-0_6,

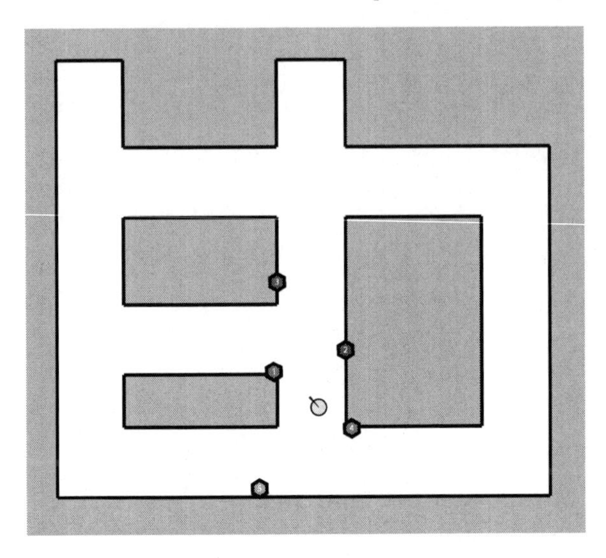

Fig. 6.1: A robot in a simulated office environment with some landmarks around it

6.1.1 A Goal-Directed Robot Navigation Task

The problem to consider is the following: An autonomous robot is requested to find a certain location in a simplified, simulated office environment (see Fig. 6.1). The environment consists of walls, represented by line segments, and landmarks the robot can perceive when they are in its field of vision and not occluded by a wall. At the start of the learning task the world is completely unknown to the agent—no map is given and no other information is provided.

A number of primitive actions (a subset of going straight, going back, and making various turns) can be taken by the robot, but it has no idea of the implications of its actions, that is, no model of the control task is given. This implies that for learning in this scenario, a model-free algorithm like Q-learning (Sect. 2.5.3) must be applied to solve the problem. No built-in collision avoidance or any other navigational intelligence is provided. The robot is assumed to be able to perceive walls around it within a certain maximum range. Also, it is supposed to be capable of recognizing unique landmarks around it.

The goal of the agent is to "find" a certain location within the environment and drive towards it. It is assumed that the agent knows when it reaches this location.

The given scenario can be formalized as a Markov decision process (MDP) $\langle \mathcal{S}, \mathcal{A}, T, R \rangle$ with

- a continuous state space $\mathcal{S} = \{(x, y, \theta) \,|\, x, y \in \mathbb{R}, \theta \in [0, 2\pi)\}$ where each system state is given by the robot's position (x, y) and orientation θ,
- an action space $\mathcal{A}$ consisting of a set of primitive actions to move the robot,
- a transition function $T : \mathcal{S} \times \mathcal{A} \times \mathcal{S} \to [0, 1]$ denoting the probability distribution that the invocation of an action a at a state s will result in a state s',

- and a reward function $R : \mathcal{S} \to \mathbb{R}$, where a positive reward will be given when a goal state $s^* \in \mathcal{S}^*$ is reached and a negative one if the agent collides with a wall.

The goal of the learning process within this MDP is to find an optimal policy $\pi^* : \mathcal{S} \to A$ that maximizes the reward the agent receives over time.

A continuous state space such as the presented one usually results in the need to use function approximation to represent the value function. Due to the inhomogeneity of the state space at the positions of walls, this approximation is crucial. Also, spatial environments are not limited in size, so the state space can become arbitrarily large. Thus, the given example describes a complex state space, making it a candidate for applying the methods derived within this work.

In the following we derive a structure space aspectualizable representation tailored for the application of reinforcement learning in the described scenario.

6.1.2 Identifying Task and Structure Space

As described in Sect. 5.2.1, any problem has two aspects that establish goal-directed and generally sensible behavior, or, put differently, consists of task space and structure space. Structure space refers to the part of the problem that remains the same over structurally similar tasks.

Example 5.2 (p. 70) distinguished between the two aspects for an office environment: Structure space refers to anything that makes the agent navigate safely through corridors, while task space is concerned with knowing where to be and where to go within the world.

In the following, we briefly investigate different ways to encode such knowledge for the current task, and then develop representations for both task and structure space.

6.1.3 Representation and Frame of Reference

The state space representation (x, y, θ) of the original MDP as described in Sect. 6.1.1 is not suitable for solving the problem with reinforcement learning because it is dependent not on the agent, but on an absolute coordinate frame. As pointed out before, that is not suitable for achieving generalization while learning.

Thus, an important consideration for representing spatial knowledge is the question of the *frame of reference* that is used, that is, how objects in the scene are aligned. Frames of reference have been extensively studied in the field of spatial cognition, see for instance Klatzky (1998) for a thorough discussion or Musto et al. (1999) for work on reference frames in the context of motion representation.

According to Levinson (2003), one can distinguish between an *allocentric* and an *egocentric* frame of reference. The distinction between them is based on the axes to which the spatial locations are aligned. In the allocentric case, these axes are given a priori and are immutable; in the egocentric case they change continuously while the agent moves. A representation is called allocentric or egocentric if it uses an allocentric or egocentric frame of reference, respectively.

An allocentric frame of reference is aligned towards an absolute orientation, such as the cardinal directions. A map is usually aligned such that north is always at the top. This is called an *absolute* frame of reference. If the position of an agent within the world is represented by its absolute coordinates (x, y) with regard to some coordinate system, this is an allocentric representation.

For achieving generalizing policies that can be reused, it is preferable to choose an egocentric frame of reference. This kind of representation describes the agent's view of the world, and it is the agent that has to make decisions in these tasks after all. The effect of these actions will again affect the agent. One could say that a frame of reference that is egocentric to the acting entity ensures that the actor is always at the center of attention, and circumstances that are important for action selection are observed there. Egocentric representations are invariant to both turns and translations in an allocentric reference system and thus are a good choice for a representation aiming at knowledge reuse.

6.2 Representing Task Space

In this section we investigate a proper representation for task space in the goal-directed robot navigation scenario.[1] As mentioned above, this is about specifying the robot's position within the environment, and the frame of reference chosen has to be an egocentric one.

6.2.1 Usage of Landmarks

When representing a position within an environment in an egocentric frame of reference, the concept of *landmarks* becomes important. Presson and Montello (1988) give the intriguingly simple definition that anything that sticks out from the background can be used as a landmark. These landmarks can then be used for navigation purposes, as the following definition by Sorrows and Hirtle (1999) clarifies: "Landmarks are prominent, identifying features in an environment, which provide an observer or user of a space with a means for locating oneself and establishing goals."

Landmarks are among the most important concepts for human spatial navigation because humans have no exact built-in localization abilities. Landmarks can be used to identify places and decision points. In robot navigation, landmarks are important especially in the context of topological mapping and navigation (for ex-

[1] Part of this section is a revised version of work published in Frommberger (2008b).

ample, Lazanas and Latombe, 1995; Owen and Nehmzow, 1998; Prescott, 1996) to describe distinguishable places within the environment.

Following Presson and Montello, anything can be used as a landmark, as long as it can be identified. However, this identification can turn out to be non-trivial in real-life systems. Laser range finders (LRFs) alone just deliver distance information, which makes it difficult to detect distinct features in the world. The use of vision systems is suitable for landmark detection, but camera-based landmark detection is heavily influenced by lighting conditions and is difficult to apply reliably. In many cases, landmark-based navigation is implemented by using artificial landmarks such as color blobs distributed over the environment.

This book does not focus on how to detect landmarks—this is a research field on its own. It is assumed that landmarks *exist* in the environments the agent operates in and that the agent is able to *detect* them with a certain reliability—in fact, we will see later that for the presented representation this reliability does not need to be high at all.

In the following, we take a look at two different approaches to include landmark information in the proposed setting.

6.2.2 Landmarks and Ordering Information

To encode positional information with the help of detected landmarks, qualitative approaches play a prominent role. As argued in Sect. 4.6, those representations are especially valuable when the size of the state space plays an important role and when taking an action-oriented view on control problems. In the following, we take a look at related work using such representations in agent control processes.

The first working approach to robot navigation that totally omitted metrical measurement and just considered qualitative environment information was the QUALNAV algorithm (Levitt and Lawton, 1990). It used cyclic ordering information of detected landmarks to approximately encode the robot's position. Even if the authors provided a working system, their claims about the regions in which the robot is supposed to be under observation of a certain spatial configuration turned out not to be entirely correct, as pointed out by Schlieder (1991); a surround view of landmarks cannot provide a unique mapping to the regions in the plane that emerge when drawing lines from each landmark to all others.

To cure the problems detected in the QUALNAV algorithm, Schlieder presented the so-called *panorama* (Schlieder, 1993). The panorama consists of the circular order of detected landmarks plus a set of *virtual landmarks* that arise by point reflection of each landmark with regard to the center (the position of the observer). This is sufficient to gain the desired unique mapping to a certain region. However, computing the panorama representation requires metrical sensory information with regard to the angles, because the projected virtual landmarks are not perceived, but have to be calculated.[2] So even while being a qualitative representation, it loses one of the strengths qualitative representations usually show: that they can be built without exact metrical information.

Schlieder's panorama has been developed further to the *extended panorama* approach (Wagner and Huebner, 2005; Wagner, 2006) that has been successfully applied in the context of robot soccer.

6.2.3 Representing Singular Landmarks

Whenever landmarks that represent an object are mentioned within this section, they are considered to be point-based. That is, we expect a landmark to be represented as a singular point in the plane that can be associated with a unique coordinate $(x,y) \in \mathbb{R}^2$. Of course, this view is idealized. But under the assumption of perfect object detection it is possible to derive such a unique coordinate when we regard the center of gravity of the object as representing its location. Abstraction of extended objects to points is a common principle in qualitative spatial approaches.

Invisible objects may also serve as landmarks. For example, they could also be RFID tags distributed all over the scene. In modern office environments this is often the case, and also in supermarkets every product nowadays can be identified by an RFID tag. The assumption of point-based landmarks is very suitable in this case.

At any point in time, the agent is surrounded by a varying number of detected landmarks within the viewing range of its sensors. Each landmark b has a certain distance d_b from the agent and an angle ϕ_b with its moving direction. The sequence $(d_1,\phi_1),\ldots,(d_n,\phi_n)$ of n detected landmarks exactly describes the position of the agent in an egocentric frame of reference. As long as $n > 1$, every sequence $(d_1,\phi_1),\ldots,(d_n,\phi_n)$ maps to exactly one position (x,y,θ) of the agent within the world, with (x,y) being the global coordinates of the robot and θ its orientation.

A representation like this fulfills the requirement of an egocentric representation. However, with regard to the state space size, there is no improvement towards a coordinate-based representation. The state space is still continuous, and very similar states (that is, positions that are very close to each other) still have different, but overly exact representations with regard to state space size.

To abstract from these differences we now investigate a qualitative representation that roughly encodes the whereabouts of landmarks with respect to the agent and its moving direction. First of all, let us abstract from the spatial extent of the robot and represent the agent by a point in the plane. Around the point representing the robot the space is partitioned into circular sectors (see Fig. 6.2). The choice of partitions can be varied. They can span the whole range of 360° or just a part of it, and they may have equal or unequal angular size. For example, the space at the back

[2] In particular, for constructing the virtual landmarks it is sufficient to be able to decide whether the angle between two landmarks is greater or less than 180° (Röhrig, 1998). While this may be comparably easy to decide for a human, a computational system needs angular measurements to make the distinction.

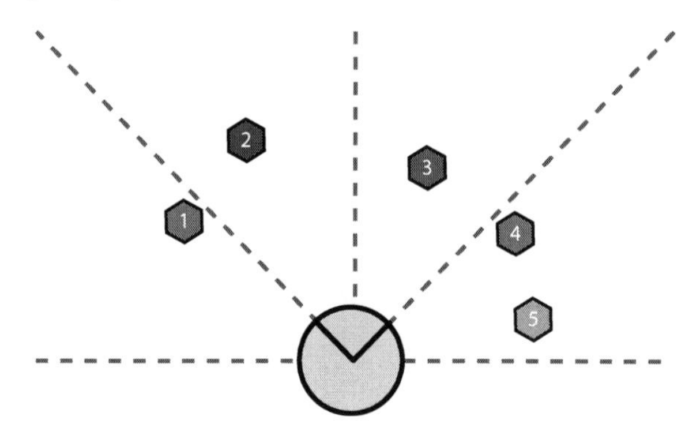

Fig. 6.2: Four sectors around the robot. Two landmarks are detected in the rightmost sector, and one each in all the others

of the robot may remain unconsidered depending on sensor capabilities or motion dynamics and action primitives of the robot.

Using such a partition is a common approach in the area of qualitative spatial reasoning (QSR). Partitions like this can be found in several spatial calculi, for instance, the "cone-based cardinal direction calculus" (Frank, 1991) that partitions the plane in an absolute frame of reference into eight sectors corresponding to the cardinal directions, or the more general star calculus (Renz and Mitra, 2004) that offers sectors in variable sizes. Such an adaptable granularity can also be found in the relative oriented point reasoning algebra $\mathcal{OPRA}_m$ (Moratz et al., 2005; Moratz, 2006). This kind of partition has also been used in a multi-robot context (Busquets et al., 2002, see Sect. 3.4.4).

Every landmark that is detected by the sensory system of the agent can now be mapped to exactly one sector in the plane (or none, if it is detected in the back). Of course, landmark positions exactly on the border between two sectors have to be mapped to exactly one sector; this mapping has to be well-defined and must be consistent such that each position always maps to the same sector for a given robot orientation.

The mapping from angles to landmarks to sectors is a coarsening. The values of the angles are regarded in rough categories instead of exact numbers. This contrasts with the panorama, where exact values are not represented, but needed to computationally build the representation. In the sector-based approach presented here the exact ordering information is not and does not need to be provided. It could also operate on rough estimates of the landmark's position, which could be, for example, derived from a vision image.

6.2.3.1 Selecting from Multiple Landmarks

When using a partition as described in the previous section one ends up with a set of landmarks for each sector. The size of each of these sets corresponds to the number

Algorithm 3 Creating a sector-based landmark representation

for $i = 1 \dots n$ **do**
  $L_i := \emptyset$
end for
Determine set L^* of detected landmarks
repeat
  Take one element $l \in L^*$
  for $i = 1 \dots n$ **do** ▷ check all n sectors S_i
    if l lies within sector S_i **then**
      $L_i := L_i \cup \{l\}$ ▷ add l to the corresponding list
    end if
  end for
  $L^* := L^* \setminus \{l\}$
until $L^* = \emptyset$

of landmarks detected in this region. Let us say that we partition the world into $n \in \mathbb{N}$ sectors and k_i $(1 \leq i \leq n)$ landmarks are detected in each sector i, $k_i \geq 0$. That creates a sequence of n sets to form the task space observation $\psi_T(s)$:

$$\psi_T(s) = (L_1, L_2, \dots, L_n), \quad L_i = \{l_{i_1}, l_{i_2}, \dots, l_{i_{k_i}}\} \tag{6.1}$$

with l_{i_j} being the detected landmarks in sector L_i. For the case of $k_i = 0$ (that is, no landmark is detected in sector i), it holds that $L_i = \emptyset$. Note that each landmark can map to exactly one sector, such that $l_{i_j} = l_{i'_{j'}}$ only holds if $i = i'$ and $j = j'$.

Algorithm 3 shows the procedure of mapping a set of detected landmarks L^* to the lists L_i. The algorithm has a runtime complexity of $O(n \times |L^*|)$; put differently, its runtime is linear in the number of landmarks detected.

In theory, a set L_i can become arbitrarily large. In malicious environments it may contain all landmarks available in the world, but even in "normal" scenarios it can happen that the number of landmarks in one sector is higher than just a few. In the example scenario of the supermarket with RFID tags as landmarks mentioned above, the sets L_i would consist of hundreds of entities such that very rarely would two observations be identical. This would prevent the system from being able to learn.

In practical use, it can become unhandy not to know how many landmarks are to be expected in such a set, for instance, when storing a Q-value for a certain representation. While this can be achieved with a hash table for example, it will be cumbersome with artificial neural networks that require a fixed set of input neurons. Furthermore, the state representation can become arbitrarily complex and hard to parse.

However, the biggest problem of input vectors of arbitrary lengths is that one ends up with an arbitrarily large number of different state representations for very similar situations—which is exactly what should be avoided, as this blows up the state space dramatically. For example, take a look at Fig. 6.3; the environment in

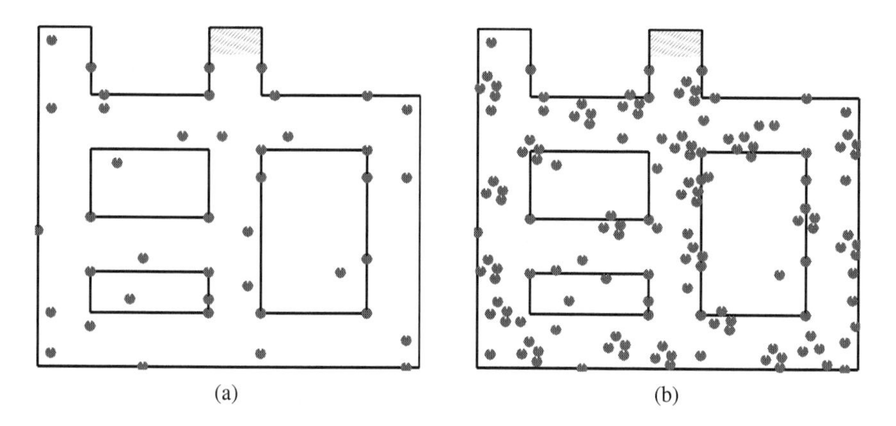

Fig. 6.3: Varying number of landmarks in an indoor environment: The number of landmarks in (a) seems to be sufficient for orientation and for finding the marked goal area, while (b) definitely bears more landmark information than needed

Fig. 6.3b provides definitely more landmark information than is needed, while it seems that the world in Fig. 6.3a contains enough landmarks for successful localization: Given an appropriate scan range, two landmarks are visible from any position. This assumption will be confirmed by an empirical evaluation in Sect. 7.2.6. So the claim is that a certain maximum number of landmarks in a sector is sufficient for encoding the position of the agent. In other words: The solution to the problem of input vectors of arbitrary lengths is, again, abstraction.

In an MDP, a selection of features violates the Markov assumption. However, we are already operating on a partially observable Markov decision process (POMDP) in the context of this work. If we ignore some of the perceived landmarks, the resulting observation will correspond to more than one state of the original state space. Then, these original states cannot be distinguished anymore. In other words, this procedure decreases the observability of the state space to a certain extent. But this kind of abstraction is exactly within the aims of the work presented here.

So the desired procedure is a set of aspectualizations within the sets L_i to achieve smaller sets $L'_i \subseteq L_i$. We chose a parameter $l_{\max} \in \mathbb{N}$ denoting a maximum number of landmarks allowed to be represented within a sector such that $|L'_i| \leq |L_i|$ and $L'_i \leq l_{\max}$ for all sets L_i.

It has to be noted that an upper bound of landmarks within a sector does not necessarily shrink the *observed* state space, that is, the set of observations really made during the learning process. In regular environments with perfect sensor information, it can even increase it. However, under occlusion and weak sensory information and with a sensible strategy to select landmarks, one experiences both a reduction of the observed state space as well as a benefit from generalization, because different original observations can share the same reduced one.

Empirical results on navigation with respect to point-based landmarks and the effect of the choice of l_{max} can be found in Sect. 7.2.6.

6.2.3.2 Landmark Selection Based on Distance

The question now is how to realize this aspectualization, that is, which landmarks to choose and which to drop from an original set L_i. We call the sequence $L_1,\dots,L_n$ the *original observation* and $L'_1,\dots,L'_n$ the *reduced observation*.

Choosing landmarks randomly is not a suitable strategy, because this results in arbitrary combinations of selected landmarks and even enlarges the observation space. An alternative is to prefer those with the smallest distance from the robot. This is based on the assumption that landmarks near the agent may have greater importance for the current situation, and, of course, the observation is more exact because the sectors' widths are smaller near the robot. The relevance of landmarks in landmark selection depends on the location of the robot.

An advantage of this strategy is that it establishes a unique mapping within the observation function. With perfect sensors assumed and $n > 1$ it holds that each position of the robot maps to exactly one reduced observation:

$$\forall (x,y,\theta)\, \exists_1\, L'_1,\dots,L'_n. \tag{6.2}$$

The number of landmarks to regard, l_{max}, can be chosen very small. Experimental results described in Chap. 7 show that in the robot navigation scenario $l_{max} = 1$ suffices for successful learning and shows better performance than higher values of l_{max}.

6.2.3.3 Situation-Dependent Choice of Landmarks

The need for landmark information is not equally high at every place within the environment. For example, in a corridor, there is only one option for a navigating robot, that is, to drive forward (assume it is not capable of performing a U-turn and stopping is not a reasonable option for the task at hand). In this case, no task space information is needed, because the structural information constituting structure space suffices to make the right decision. The insight that landmarks are the most important at decision points and less useful in between is a well-known fact in human wayfinding research (see Richter, 2007, for example). Humans also use landmarks for reaffirming that they are still on the right track; however, in non-decision states (see Sect. 5.6.1), this does not affect action selection.

Thus, structure-induced task space aspectualization (SITSA) as introduced in Sect. 5.6 can successfully be applied to robot navigation tasks. Non-decision states can be identified by analyzing policies of earlier learned tasks and used to abstain

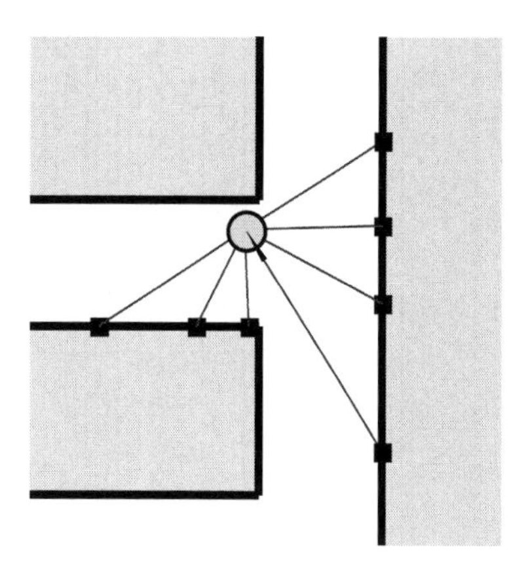

Fig. 6.4 Sample scanning of walls serving as extended landmarks: The walls are only perceived as a sequence of angles relative to the agent. This can be seen as landmark scans (lines) hitting obstacles (small boxes)

from regarding landmarks at these places. This procedure to select landmarks by the use of SITSA is called *structural-decision-aware landmark selection* (SDALS).[3]

Empirical results on the use of SITSA for landmark selection are given in Sect. 7.2.7.

6.2.4 Views as Landmark Information

In contrast to considering point-based landmarks, we now take a look at another, more abstract way to represent landmark information around the agent. Instead of searching for landmarks in the scene and classifying them according to their position, let us take a look at the *view* around the agent. It is assumed to have rich sensory information of the robot's surroundings. Navigation approaches based on views have been implemented, for example, within the *view graph* approach (Mallot et al., 1997; Franz et al., 1998).

To represent the view in a manageable representation, samples are taken around the robot by checking the environment at a predefined set of angles. Instead of having point-based landmarks, it is now assumed to have extended objects that serve as landmarks in the scene. With those objects arranged around the agent, there is at most one landmark that the agent can perceive when "looking" along a line of a certain angle: the object next to it in the direction of sight that occludes more distant objects (see Fig. 6.4).

6.2.4.1 Walls as Landmarks – the Realator Approach

In an office environment, the agent is usually surrounded by walls, so walls are a good choice of landmarks, assuming that they can be uniquely identified. A promi-

[3] First ideas on this landmark selection procedure were published in Frommberger (2008b).

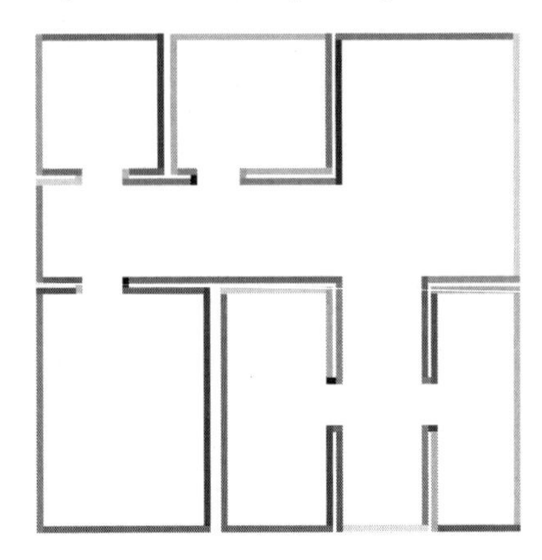

Fig. 6.5 The Realator: Each wall is given an individual color and is uniquely recognizable

nent model of such a setting is the *realator* (Barkowsky et al., 1994, 1995). The realator models an indoor environment in which each wall is uniquely distinguishable by its color (see Fig. 6.5). In this setup a qualitative navigation strategy was demonstrated which needs the a priori knowledge of the circular sequence of the colored walls in the scene and a separate obstacle avoidance mechanism, but does not rely on distance information (Röhrig, 1998). The continuous view of colors around the robot serves as the agent's input.

In this work, we use a similar representation, with the difference that the whole view of colors is not regarded, but, as mentioned before, sampled at distinct angles around the robot. We refer to this as the *sampled view model.*

6.2.4.2 Sampling Landmark Information Around the Agent

To facilitate readability, we assume without loss of generality the traveling direction of the agent to be an angle of 0°. There are n angles with regard to this direction, $\beta_1,\ldots,\beta_n$. Ideally, these samples are chosen symmetrically to the moving direction, that is, $\beta_i = -\beta_{n+1-i}$ for $1 \leq i \leq n$. Whether the samples should cover a whole 360° view or are restricted to the front view of the robot (for example with $\beta_1 = -\frac{\pi}{2}$) is a matter of design.

At each angle, the agent perceives a wall (or no wall, if there is none within its viewing range). One special symbol is reserved for the perception of no wall (0 is an appropriate choice), and all other walls have a unique number. So the task space observation ψ_T of a state $s \in \mathcal{S}$ is

$$\psi_T(s) = (c_1,\ldots,c_n) \tag{6.3}$$

with $c_i \in \mathbb{N}_0$ being the identifier of the wall detected at angle β_i. In the realator-inspired framework, $(c_1,\ldots,c_n)$ is a sequence of colors detected around the agent.

Figure 6.6 depicts which physical positions in the original state space $\mathcal{S}$ correspond to the same landmark-based representation $\psi_T(s)$ under the sampled view

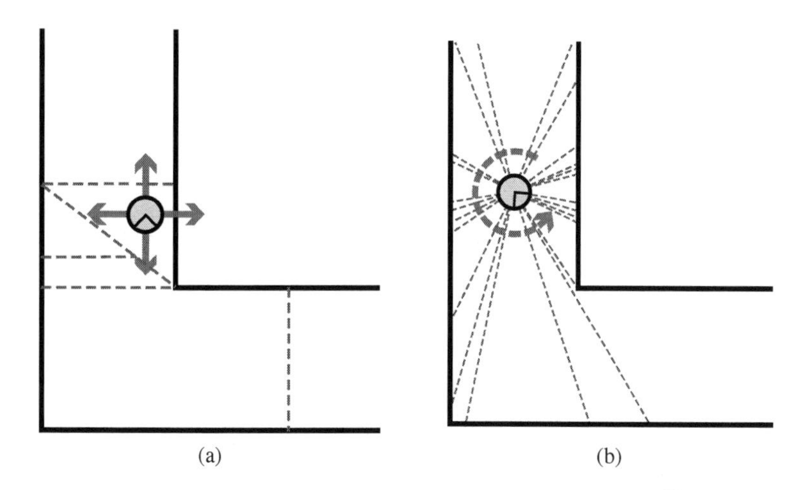

Fig. 6.6: Perceptual regions created with the sampled view model. The dotted lines are borders of regions with the same observation of landmarks under translation (a) and rotation (b) of the robot. While the robot is in any of the regions in (a), the sampled landmark view does not change as long as it does not change orientation. In (b), the regions denote angular intervals of orientations that share the same landmark view when the robot maintains in its position. The regions refer to five landmark scans around the robot at $-90°$, $-45°$, $0°$, $45°$ and $90°$

model. The agent is moved, but not turned (Fig. 6.6a), or turned, but not moved (Fig. 6.6b). Under translation (Fig. 6.6a), regions with identical observation are rather large, while the angular intervals that share the same observation (Fig. 6.6b) generate a high resolution.

6.2.4.3 Benefits of the Sampled View Method

The assumption that every wall can be uniquely identified (as symbolized by the unique color coding) is not very realistic. However, for most experiments in this work this model is used for demonstration in the simulator.

The first reason for the choice of the sampled view model is that the detection of a colored wall is easily comprehensible when looking at the simulator and the sequence of colors, so the perception of the robot is evident and can easily be traced. A sequence of colors is eye-catching and can nicely be presented, and the corresponding walls can be found fast within the world. In contrast, the identifiers of point-based landmarks do not offer such a bold representation of task space for the viewer.

The second, more important reason, is that the number and locations of point-based landmarks within the world critically influence the learning speed (see experiments on this issue in Sect. 7.2.6). When using the walls for task space representation, the number of landmarks scales with the size of the world: A structurally larger and more complex world also bears an increased number of landmarks. In the sampled view model, the learning performance does not rely on the distribution of landmarks over the scene, as this distribution is incorruptibly induced by the environment's structure. Thus, a sampled view representation is much better suited for evaluation. However, all presented methods work with either point-based landmarks or sampled views.

6.2.5 Navigation Based on Landmark Information Only

The task space representation based on landmarks as presented in the previous section is already a very rich description of the world. In this section, we take a look at whether this description alone can be used for solving the goal-directed navigation task. Especially interesting is the question of what this representation encodes and what it does not.

6.2.5.1 Perceptual Aliasing

The encoding of a circular order of perceived colors is sufficient to approximately represent the position of the agent within the world and to derive a sequence of actions to reach the goal state. However, it is not sufficient to prevent the robot from collisions. As stated above, the mapping from physical locations to the state representation is not unique, and given the same system input, the consequences of an applied action can differ dramatically and prevent stable learning. This problem is known as *perceptual aliasing* (Whitehead and Ballard, 1991; Whitehead, 1992).

Practically speaking, every environment is partially observable: Real-world systems are always susceptible to perceptual aliasing, no matter what sensors they are equipped with. In this special case, it is not a question of sensory shortcomings, but a consequence of the desire to keep the system's input as small as possible—the less comprehensive the input, the more the danger of experiencing perceptual aliasing.

A thorough study by Crook and Hayes (2003) shows that Q-learning with 1-step-backups (see Sect. 2.5.2) performs very badly in situations with perceptual aliasing, but with n-step-backups, that is, with the usage of eligibility traces (Sect. 2.5.2), a stable policy can be learned in certain cases.

6.2.5.2 Perceptual Aliasing in the Robot Navigation Task

In the scenarios presented in this work the use of eligibility traces alone with a pure task space representation usually results in unstable learning behavior with a high risk of collisions because the same state-action at the same system state will

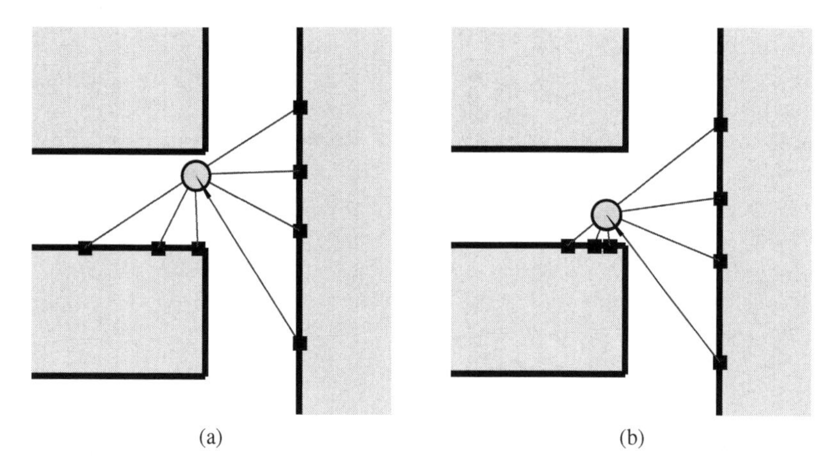

Fig. 6.7: Perceptual aliasing: the same state representation, but different consequences. While receiving the same sequence of detected walls as sensory input, situation (b) will result in a collision when moving forward at the next time step. In situation (a), no collision will occur

sometimes result in a collision and sometimes not (see Fig. 6.7). Navigating too closely around corners results in the shortest path towards the target, but it comes at the high risk of a possible collision, as the robot is not an abstract point, but has a certain extent. So the same action pair is sometimes positively rewarded in one episode and negatively rewarded in another.

Summed up, a pure task space representation encodes the agent's position quite well in our scenario, but does not encode any information about the agent's relation with regard to the obstacles, and it does not take the structure of the world into account by ignoring the positions of walls. The same problem is reported for the topological approach by Lane and Wilson (2005, see Sect. 3.4.4), which also cannot cope with inhomogeneous state spaces like the one of the task at hand. State space abstraction without regarding structural state space information hinders the agent from developing a generally sensible spatial behavior. Thus, we need to represent structure within the state space.

Empirical results on navigation only based on landmark information are described in Sect. 7.2.5.

6.3 Representing Structure Space

In the following, we develop a representation to encode structure space. Representing structure space in the robot navigation scenario means explicating the structure of the environment. As pointed out in Example 5.1 the structure of the world is induced by walls. Walls form corridors, rooms, and open spaces, and by this they induce a generally sensible behavior for an agent operating in this environment. The whereabouts of the walls are important with respect to the agent's orientation, that is, its moving direction. This information forms structure space in this scenario.

Walls can be abstracted to line segments, as is done, for example, in architectural floor plans. The simulator used in this work also makes use of this 2-D projection. Taking the visibility range into account, the robot's observation of the walls can be regarded as a set of non-overlapping line segments.

The following section describes the *relative line position representation (RLPR)* which encodes the configuration of line segments around the robot to represent structure space in an office environment.[4]

6.3.1 Relative Line Position Representation (RLPR)

Again, as the position of the wall is important with respect to the agent's orientation, we choose an egocentric frame of reference.

The different instances of this representation should be differentiated between according to appropriate action selection (see Sect. 4.4.1). In particular, the exact angle of the line segment with respect to the agent's orientation does not matter as long as the action to take in this situation does not change. In this case, no decision boundary is crossed. The information that the agent is within a corridor will cause actions to follow this corridor, regardless of whether the walls are completely straight or parallel (see Fig. 6.8). Thus, we do not encode the concrete position of the wall; instead we choose a qualitative description of the relation of the line segment to the robot.

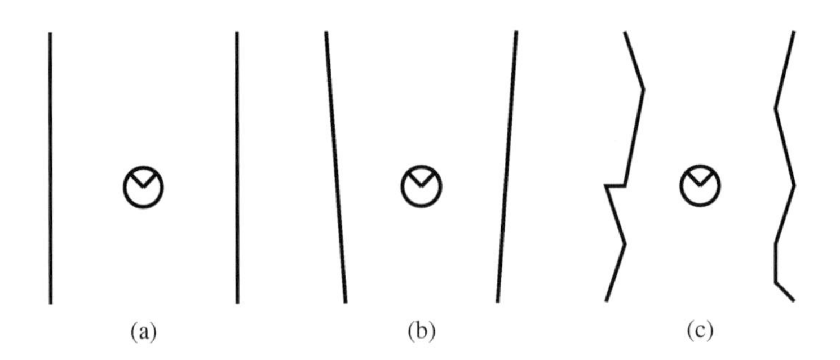

Fig. 6.8: Different layouts of corridors with identical consequences: Going straight forward is the desirable action for the robot

[4] RLPR has been introduced in Frommberger (2006). Parts of this section originate from this paper or from follow-up publications (Frommberger, 2007b, 2008a).

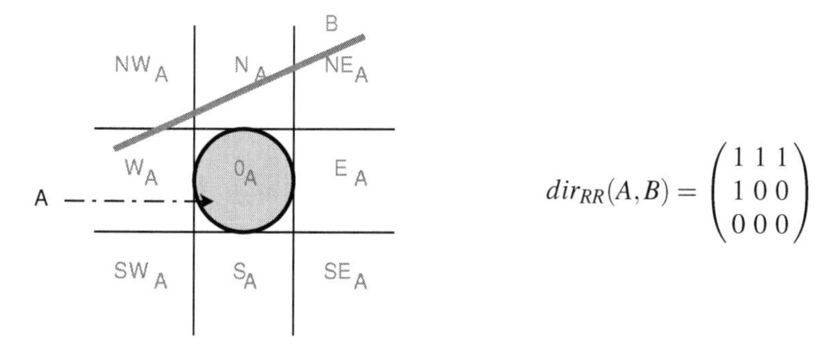

$$dir_{RR}(A,B) = \begin{pmatrix} 1 & 1 & 1 \\ 1 & 0 & 0 \\ 0 & 0 & 0 \end{pmatrix}$$

Fig. 6.9: Partition of the direction-relation matrix around a circular object A and the matrix $dir_{RR}(A,B)$ for a line B with respect to A. "1" stands for "true" and "0" stands for "false"

The robot itself is not an abstract point, but has a certain extent. Its size is important for action selection, because the agent is not supposed to perform actions that bring it into situations where it collides because of its size, for example, passing through too-narrow gaps.

We are aiming at representation that helps describe the relations of an extended object (the agent) to other extended objects (the line segments). A suitable framework to represent such relations is the *direction-relation matrix* (Goyal and Egenhofer, 2000; Goyal, 2000).

6.3.1.1 The Direction-Relation Matrix

The direction-relation matrix is a means to describe the relations between two extended objects A and B in an allocentric frame of reference. It is a partition of 2-D space that instantiates the *projection-based directions* introduced by Frank (1996). Four lines divide the space around the extent of a reference object A to form a 3×3 grid around it. Motivated by geo-information sciences, the grid cells are labeled according to the cardinal directions: N_A for north of A, NE_A for north-east of A, and so on. 0_A is called the *neutral zone* in Frank (1996). It is the central grid cell where the reference object A is located, such that 0_A is the bounding box of A. Apart from 0_A, all grid cells have an open extent, that is, one or two of their boundaries lie in infinity.

The direction-relation matrix $dir_{RR}(A,B)$ is defined as follows

$$dir_{RR}(A,B) = \begin{pmatrix} NW_A \cap B & N_A \cap B & NE_A \cap B \\ W_A \cap B & 0_A \cap B & E_A \cap B \\ SW_A \cap B & S_A \cap B & SE_A \cap B \end{pmatrix}. \tag{6.4}$$

$NW_A \cap B$ defines a boolean value that denotes whether B intersects NW_A, that is, whether a part of object B is located in grid NW_A. The other boolean values are defined accordingly. Figure 6.9 shows an example with a circle as a reference object and a line.

A spatial partition like this does not necessarily need to be used with an allocentric frame of reference. If object A has an intrinsic front or a clearly defined moving direction, the grid can also be constructed relative to the orientation of A. This establishes an egocentric frame of reference. In Mukerjee and Joe (1990), such an egocentric grid has already been used to model the movement of extended objects in a qualitative manner. In a similar way, we now use this partition to encode structural information of the local surroundings of a moving robot.

An important property of this partition is that the regions in the front, at the back, and at the sides of the moving object have a width based on the agent's size. An interpretation that can be drawn from this fact is that if there is no other object detected within this region (if, for example, $\mathrm{N}_A = 0$ in $dir_{RR}(A,B)$), the object A has sufficient space to move there.

6.3.1.2 Line Configurations in the RLPR Grid

To model structural spatial information in the relative line position representation (RLPR), we partition the space around the agent (the reference object) in the same way the direction-relation matrix does, but with an egocentric reference frame based on the agent's front. In the following, we refer to this partition as the *RLPR grid*. The resulting regions are enumerated with R_0 corresponding to the neutral zone and R_1 to R_8 for the other grid cells starting with R_1 for SE_A and then moving counter-clockwise to R_8, which corresponds to S_A (see Fig. 6.10). The set $\mathcal{R} = \{R_0, \ldots, R_8\}$ defines the RLPR grid.

The size of the grid defining the immediate surroundings is given a priori. It is a property of the agent and depends on its size and system dynamics (for example, the robot's maximum speed).

Now let us look at the whereabouts of the walls, abstracted to n line segments B_i, within the RLPR grid. $\mathcal{B} = \{B_1, \ldots, B_n\}$ is the set of line segments detected in the scene. The first question for representing the position of line segments within the scene is "Is wall B in region R_i?" The answer is encoded by the *overlap status* function $\tau : \mathcal{B} \times \mathcal{R} \rightarrow \{0,1\}$:

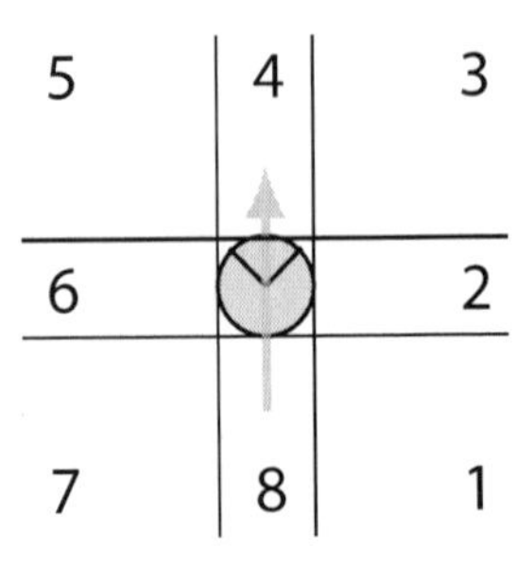

Fig. 6.10 The RLPR regions $R_1, \ldots, R_8$ arranged around the agent according to its moving direction

$$\tau(B,R) = \begin{cases} 1 & \text{if } B \cap R \neq \emptyset \\ 0 & \text{else} \end{cases} \tag{6.5}$$

with $B \in \mathcal{B}$ and $R \in \mathcal{R}$. $\tau(B,R)$ is 1 if a line B cuts region R and 0 if not.

This overlap status has to be regarded for the whole scene. The overall number of lines in a region $R \in \mathcal{R}$ therefore is defined by a function $\overline{\tau} : \mathcal{R} \to \mathbb{N}_0$:

$$\overline{\tau}(R) = \sum_{B \in \mathcal{B}} \tau(B,R). \tag{6.6}$$

In most cases it does not matter how many line segments there are in a region—in fact, this often cannot be reliably observed due to weak sensory information. The important information is whether there are *any* lines within a region or not. So we additionally define $\overline{\overline{\tau}} : \mathcal{R} \to \{0,1\}$ as follows

$$\overline{\overline{\tau}}(R) = \begin{cases} 1 & \text{if } \overline{\tau}(R) > 0 \\ 0 & \text{else} \end{cases}. \tag{6.7}$$

The spatial structure of a scenario is given by the configuration of line segments. $\tau(B,R)$ tells us in which region a line segment B is located, but not how it is oriented. But the latter is particularly interesting for anticipatory navigation. A corridor with a left turn, for example, has connected line segments across all the right and front sectors, but none in the freely accessible space to the left. To additionally encode whether a line segment B spans from one region to another, let us determine if a line B lies within counter-clockwise adjacent regions R_i and R_{i+1} (for R_8, of course, R_1 needs consideration). The *adjacency status* $\tau' : \mathcal{B} \times \mathcal{R} \rightsquigarrow \{0,1\}$ for $i > 0$ is defined as

$$\tau'(B,R_i) = \begin{cases} \tau(B,R_i) \times \tau(B,R_{i+1}) & \text{if } 1 \leq i \leq 7 \\ \tau(B,R_i) \times \tau(B,R_1) & \text{if } i = 8 \end{cases}. \tag{6.8}$$

The overall adjacency status $\overline{\tau}' : \mathcal{R} \to \mathbb{N}_0$ and the boolean case $\overline{\overline{\tau}}' : \mathcal{R} \to \{0,1\}$ are defined analogously to τ':

$$\overline{\tau}'(R) = \sum_{B \in \mathcal{B}} \tau'(B,R), \tag{6.9}$$

$$\overline{\overline{\tau}}(R) = \begin{cases} 1 & \text{if } \overline{\tau}'(R) > 0 \\ 0 & \text{else} \end{cases}. \tag{6.10}$$

In the following, we refer to the set $\{\tau, \overline{\tau}, \overline{\overline{\tau}}, \tau', \overline{\tau}', \overline{\overline{\tau}}'\}$ as the *RLPR functions* or τ-functions.

The τ-functions describe relations in "real" space, which has spatial constraints, so the τ-functions cannot take arbitrary values. For example, the $\overline{\overline{\tau}}$ and $\overline{\overline{\tau}}'$ are interdependent. If a line segment spans from one region to another, there must be a line segment within both these regions:

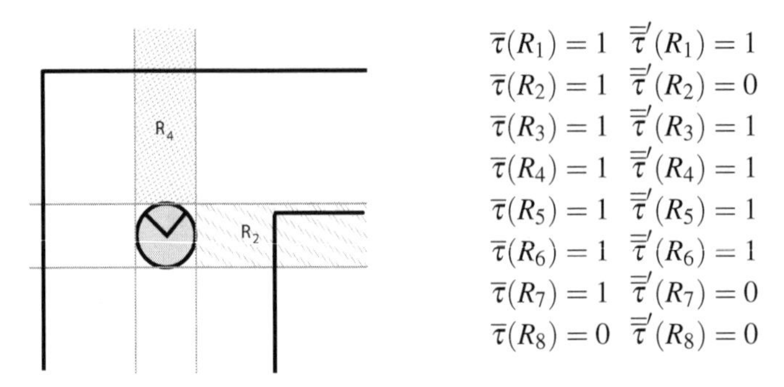

Fig. 6.11: RLPR values in an example situation. Region R_2 (right) and R_4 (front) are marked

$$\overline{\overline{\tau}}'(R_i) = 1 \quad \Rightarrow \quad \overline{\tau}(R_i) = 1 \wedge \overline{\tau}(R_{i+1}) = 1. \tag{6.11}$$

Figure 6.11 shows an example of the overlap and adjacency values of RLPR.

6.3.1.3 The Role of the Immediate Surroundings

Regarding the adjacency status, there is also the possibility of a line being in two non-adjacent regions if it cuts the neutral zone R_0, for example, if it goes from R_4 through R_0 to R_2. This can only happen if the wall is very close to the robot.

The immediate surroundings of the agent need special consideration anyway: Overlap and especially adjacency status of detected line segments as presented are interesting information to use for general orientation and mid-term planning. But the agent is supposed to react differently when detecting features in its immediate surroundings. If an object is perceived there, circumstances are encountered that make a collision possible, so an immediate consideration about actions to invoke is required. For example, a wall in front of the robot is no problem if it is away far enough, and if approaching a corridor turn, the agent always faces a wall in front of it. But if the wall is very near to the robot, driving forward should be avoided at any price. The distinction between a far and a near wall makes for a *qualitative* difference. So the immediate surroundings are important for obstacle avoidance mainly. On the other hand, information about very near objects can also be used to realize certain behaviors, for example, a wall following strategy.

To account for this issue, let us introduce another set of grid cells for the immediate neighborhood of the moving agents. On the one hand, there are the regions $R_1, \ldots, R_8$ that are limited by the perceptual capabilities of the robot only. On the other hand, bounded subsets of those regions represent the immediate surroundings (see Fig. 6.12b). The τ-functions are also defined on the regions R_9 to R_{16}. So a qualitative representation of distance is used to model the named qualitative difference: The qualitative model only distinguishes between "near" and "not near"—as for action selection only this distinction matters. Only if an object is "near" is the robot supposed to select actions to avoid it. A related spatial representation that also encodes distance qualitatively is the granular point position calculus GPCC (Moratz, 2005).

With information about obstacles in regions near the robot, let us now look again at the problem mentioned at the beginning of this paragraph that lines do not necessarily lie in adjacent regions R_1 to R_8. But if this is the case for a line segment B, then the following two statements hold:

1. B cuts the neutral zone R_0: $\tau(B,R_0) = 1$.
2. B cuts two regions in the immediate surroundings: $\exists i,j : i \neq j \wedge i,j \in [9,16] \wedge \tau(B,R_i) = \tau(B,R_j) = 1$.

Thus, the cases where lines do not cross the boundaries of adjacent RLPR regions can be identified by taking the neutral zone and its immediate surroundings into account.

The size of the regions in the immediate surroundings is specific to any agent. RLPR is a property of the agent, which means that each agent has its own RLPR grid. Depending on the motion dynamics of the agent, the concept of "near" has to be seen differently: Whether an object is near or not depends on whether actions need to be invoked immediately to avoid the emerging obstacles. If a platform is inert or fast, it may need larger immediate surroundings, and smaller ones are appropriate for a slow or agile system. The regions have to be large enough that the robot is able to invoke actions to avoid obstacles appearing in the immediate surroundings, but not significantly larger.

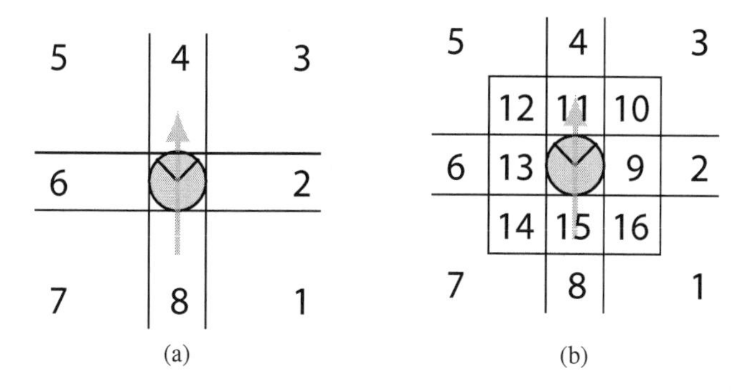

Fig. 6.12: Neighboring regions around the robot in relation to its moving direction. Note that the regions in the immediate surroundings (b) are proper subsets of $R_1,\ldots,R_8$ (a)

6.3.2 Building an RLPR Feature Vector

For regions in the immediate surroundings (R_9 to R_{16}), it is particularly important to know whether there is something within those regions or not; adjacency information is of minor importance there. Thus, only $\overline{\overline{\tau}}(R_i)$ is regarded for regions R_i with $i \geq 9$. The same holds for the neutral zone R_0. For R_1 to R_8, of course, one wants to obtain the adjacency statuses $\overline{\overline{\tau}}'$ to see where walls are leading to and thus comprehend the structure of the local surroundings of the agent. This information serves as a structure space description of the robot navigation scenario, such that the structure space observation ψ_S of state $s \in \mathcal{S}$ is defined as

$$\psi_S(s) = (\overline{\overline{\tau}}'(R_1), \ldots, \overline{\overline{\tau}}'(R_8), \overline{\overline{\tau}}(R_0), \overline{\overline{\tau}}(R_9), \ldots, \overline{\overline{\tau}}(R_{16})). \tag{6.12}$$

In the following, we refer to this representation of structure space as given in (6.12) as RLPR_∞.

Example 6.1. With ψ_S as defined above, structure space is represented as a vector of 17 boolean values, so that theoretically $2^{17} = 131,072$ structural states can be distinguished. However, many of them are physically impossible, as there are spatial constraints that lead to interdependencies between the feature values. For example, the constraint given in (6.11) reduces the number of valid configurations by a factor of 4, so that only $32,768$ possible configurations are left. Omitting certain regions (for example in the back of the agent) leads to a further reduction of possible inputs.

6.3.3 Variants of RLPR

RLPR_∞, as proposed above, is the basis representation that collects all the information that can be derived from the direction-relation-matrix-motivated grid in Fig. 6.12. However, in certain cases, not all of this information is needed.

As stated above, the RLPR grid is a property of the agent. If now, for example, an agent cannot move backwards, the information about the immediate surroundings behind is not relevant for successful action selection. Then, the values $\overline{\overline{\tau}}'(R_7)$ and $\overline{\overline{\tau}}'(R_8)$ as well as $\overline{\overline{\tau}}(R_{14})$ to $\overline{\overline{\tau}}(R_{16})$ can be omitted from the RLPR_∞ feature vector (6.12):

$$\psi_S(s) = (\overline{\overline{\tau}}'(R_1), \ldots, \overline{\overline{\tau}}'(R_8), \overline{\overline{\tau}}(R_0), \overline{\overline{\tau}}(R_9), \ldots, \overline{\overline{\tau}}(R_{13})) \tag{6.13}$$

For further reference, we call this RLPR feature set RLPR_0.[5]

Another factor is the limitation of sensory information. The RLPR_∞ grid assumes a 360° view of the world, which is rarely given. On mobile robots, there is often only a 180° laser range finder mounted to the front. In this case, even more features of

[5] RLPR_0 is explicated here, because this representation was the one originally introduced as RLPR in the first publications on RLPR (Frommberger, 2006, 2007b, 2008a). Furthermore, many experiments in this work have been performed with use of RLPR_0.

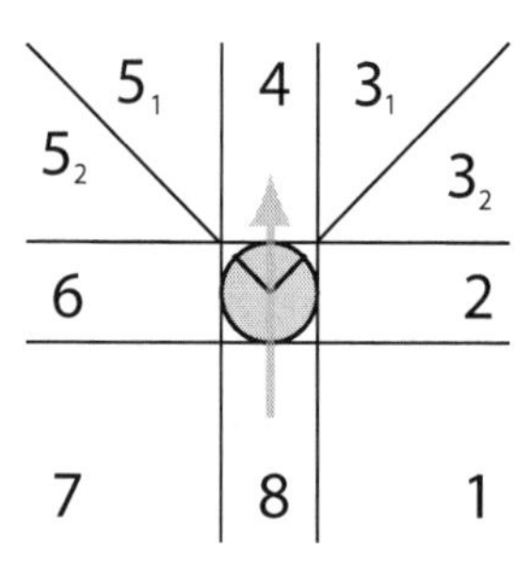

Fig. 6.13 A variation of the original RLPR grid (Fig. 6.12) that divides regions R_3 and R_5 into two parts to assure a higher resolution in this area

RLPR_∞ can be omitted, because they are constantly 0 due to the lack of sensory information in these areas.

On the other hand, variations that enhance the original RLPR grid are also possible. Figure 6.13 shows a variant (labeled RLPR_1)[6] of the RLPR partition that enhances RLPR_0. It divides the regions in the front left and right of the agent into two parts and thus increases the resolution in this area. The regions in the immediate surroundings are not divided, because this distinction only refers to mid-scale considerations.

With RLPR_1, the agent is able to detect structural changes ahead of it earlier than with RLPR_∞. For example, when approaching a turnoff in a corridor, this turnoff will only be identified when the robot has reached it. With RLPR_1, the turnoff will be identified quite some time earlier (see Fig. 6.14), so that actions to prepare a possible turn into this turnoff can be invoked early enough to ensure a smooth navigation behavior.

An empirical comparison of some RLPR variants is given in Sect. 7.2.1.2.

6.3.4 Abstraction Effects in RLPR

RLPR is a very strong abstraction from the manifold of details within the world to a comparably small number of spatial concepts. It is tailored to encode only the details relevant for reasonable action selection in indoor environments. In particular, RLPR abstracts from the following aspects:

Rotation/translation of the world. As RLPR uses an egocentric frame of reference, it is independent on the alignment of the overall environment.

Proportions. As long as the immediate surroundings are not affected and the viewing range is sufficient, RLPR abstracts from proportions of the world. The length of a corridor, for example, is not regarded in this representation.

Exact angles of corridors/turns. Angles need not be known at high resolution. A turn of 85° or 95° has the same RLPR representation in most of the agent's po-

[6] The numbering of the RLPR variants appears to be arbitrary. In fact, the numbers arise from the numbering of the RLPR modes in the simulation code. There is no semantics behind the numbers.

sitions. For example, the situations in Fig. 6.8 all share the same RLPR_∞ representation.

Irrelevant structures (gaps). Structural information that is irrelevant for navigation is ignored. Gaps between line segments that are not large enough for the robot to pass through are not taken into account for the representation.

Straight lines / curved lines. Spatial structures that are not formed by line segments (for example, curved lines) can still be approximated by lines, and the RLPR values are hardly affected by that. Such an approximation ensures that round or irregularly shaped objects can be handled by RLPR as well.

Because it is a pure structure space representation, RLPR allows for non-local generalization. Thus, RLPR is especially suitable for combining with task space tile coding (Sect. 5.3.3).

6.3.5 RLPR and Collision Avoidance

The ability not to collide with obstacles is an important aspect of generally sensible navigation behavior. However, it is important to note that RLPR covers more than collision avoidance, which is taken into account by the RLPR regions representing the immediate surroundings of the agent.

The other regions refer to mid-term planning of navigation. As can be seen in Fig. 6.14, the layout of the RLPR grid has consequences for the navigation behavior of the agent: In a finer representation, actions to enable driving around a corner can be invoked much earlier. So the τ-functions operating on the outer regions of

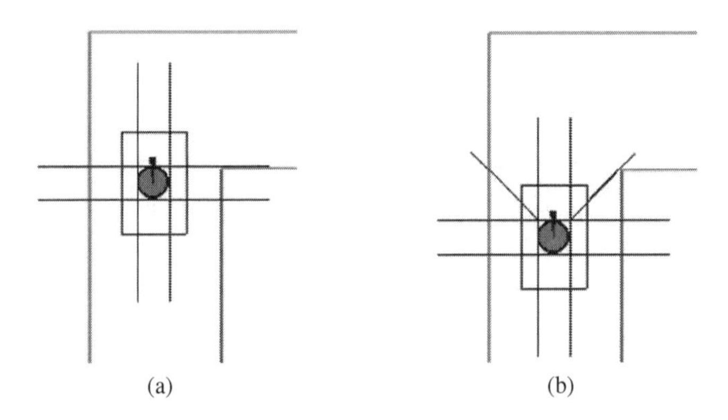

Fig. 6.14: Comparison of RLPR_∞ (a) and RLPR_1 (b). Both situations depict the moment when the agent notices the end of the right wall according to the RLPR values. With RLPR_1 the turn is detected significantly earlier than with RLPR_∞. Appropriate action to approach the curve can be invoked in time

RLPR detect the structures the agent is operating in and allow for an anticipatory navigation that goes way beyond collision avoidance.

6.4 Landmark-Enriched RLPR

Now that abstractions are defined for both task space representation (Sect. 6.2) and structure space representation (Sect. 6.3), it is possible to combine them to achieve a structure space aspectualizable state space representation. The observation function ψ simply concatenates task and structure space observation functions:

$$\psi(s) = (\psi_T(s), \psi_S(s)). \tag{6.14}$$

This combination of landmark and RLPR information is called *landmark-enriched RLPR* or, abbreviated, *le-RLPR*.

For example, an le-RLPR observation function using the wall colors as landmarks with seven samples for task space representation (6.3) and RLPR_0 as structure space representation (6.13) looks as follows

$$\psi(s) = (c_1, \ldots, c_7, \overline{\overline{\tau}}'(R_1), \ldots, \overline{\overline{\tau}}'(R_8), \overline{\overline{\tau}}(R_0), \overline{\overline{\tau}}(R_9), \ldots, \overline{\overline{\tau}}(R_{13})). \tag{6.15}$$

Figure 6.15 depicts the flow of abstraction and the used abstraction principles when building and accessing le-RLPR.

6.4.1 Properties of le-RLPR

In this section we investigate how well the observation space created by le-RLPR matches the criteria for efficient abstraction formulated in Sect. 4.5.4: significant size reduction, discrete state space, high π^*-preservation quota, and high accessibility.

High accessibility. By design, le-RLPR is aspectualizable with regard to task and structure space, because it concatenates a task space representation and a structure space representation. Both task and structure space descriptors can be achieved by an aspectualization (see Fig. 6.15). Thus, high accessibility is given.

Discrete state space. The enumeration of landmarks as well as the result of the τ-functions are discrete. Thus, Image(ψ) is discrete, and le-RLPR creates a discrete observation space.

High π^*-preservation quota. RLPR has been designed to represent decision boundaries of the underlying state space. Landmarks assure a partition into homogeneous regions (see Fig. 6.6). Thus, in combination, a π^*-preservation quota is granted.

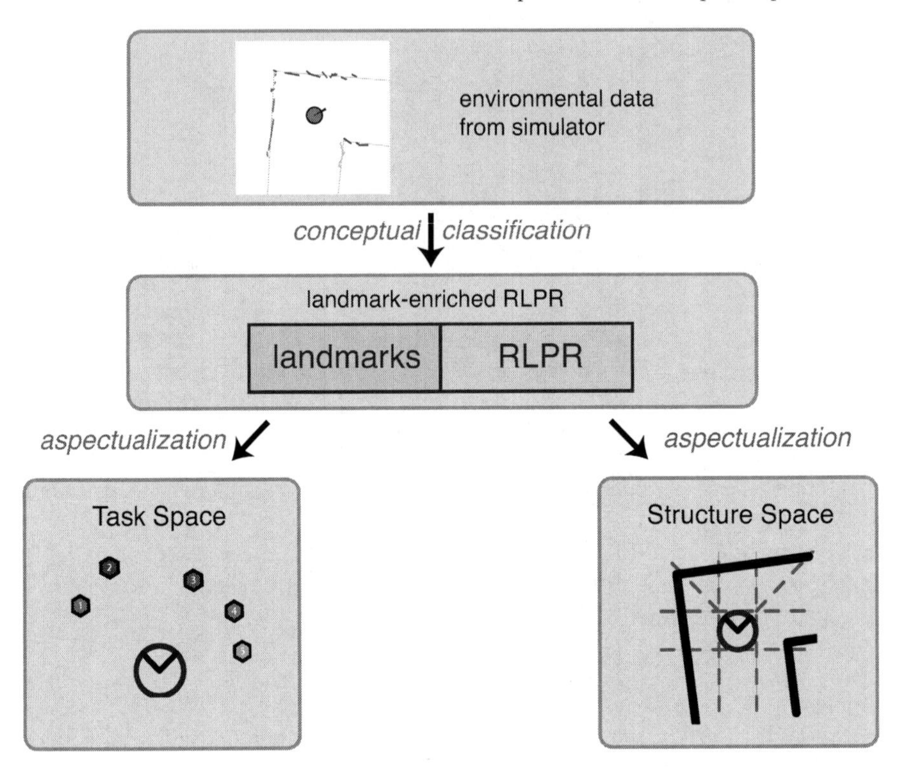

Fig. 6.15: Abstraction principles when building landmark-enriched RLPR

Significant size reduction. RLPR collapses the state space into a small and manageable number of abstract concepts. Landmark selection methods also assure a significant reduction in size for task space representation.

High generalization. le-RLPR offers a strong, yet differentiating spatial abstraction. For positional information it abstracts from the concrete position within the world by building regions according to landmark information (see Fig. 6.6). The far-reaching abstraction properties of RLPR have been pointed out in Sect. 6.3.4.

Summed up, le-RLPR fulfills the criteria for an efficient state space abstraction in an agent control context. An empirical evaluation on the performance of le-RLPR can be found in Sect. 7.2.

6.5 Robustness of le-RLPR

le-RLPR offers a high degree of abstraction of real-world information. Qualitative representations establish hard boundaries within the state space when distinguishing between concepts. While fuzzy representations aim at having continuous transitions between concepts, qualitative approaches intentionally do not provide that. In the example of RLPR, an object is "near" the agent if it is detected in the immediate surroundings, and "not near" otherwise. "Nearness" is not a graded concept here. Similarly, there is either a wall in front of the agent or not, and a landmark has been detected, or it has not. Qualitative representations make a clear statement which is easy to operate on.

Regarding the possibility of a wrong classification of a concept caused, for example, by erroneous sensory information, this clear distinction can lead to problems because any misclassification makes a *qualitative* difference. If an object that is near is classified as being far (or not existent), the consequences for the operating agent may be severe.

6.5.1 Robustness of Task Space Representation

The whereabouts of landmarks are represented quite roughly in le-RLPR. Point-based landmarks are mapped to sector regions and extended landmarks (walls) are sampled at specific places.

Three types of errors in landmark detection can be distinguished:

1. a landmark is perceived at the wrong location: The estimate of its position is affected by sensor noise or distortions caused by the agent's motion;
2. a landmark is not detected at all: It is overlooked;
3. a landmark is detected mistakenly: It is identified as another landmark in a sensor reading.

A point-based landmark being perceived at a wrong location and the sample of the extended landmark being displaced are equivalent. In the following, we only regard the point-based case.

A wrong detection of the landmark's position will often remain unnoticed, because the detected position will refer to the same region. If this is not the case, the landmark will most likely be mapped to adjacent regions. In most of the cases the differing representations then share similar semantics. Only if this is not the case—when crossing a decision boundary—the misclassification has an effect on action selection.

Because the agent makes a new decision at every new observation, a wrong task space representation will only affect *one* particular action. Thus, it can be expected that a wrong decision will be compensated for by the following action of the agent.

Furthermore, in structure space aspectualizable state spaces, misclassifications in task space are not severe. Task space knowledge does not tackle critical movements that could be dangerous for the robot—those are covered by generally sensible behavior and, thus, are a matter of structure space. Especially when using generalization methods as task space tile coding, we can expect the agent's actions to be reasonable even if landmark detection is distorted, because unknown observations then are covered by a meaningful structure space policy.

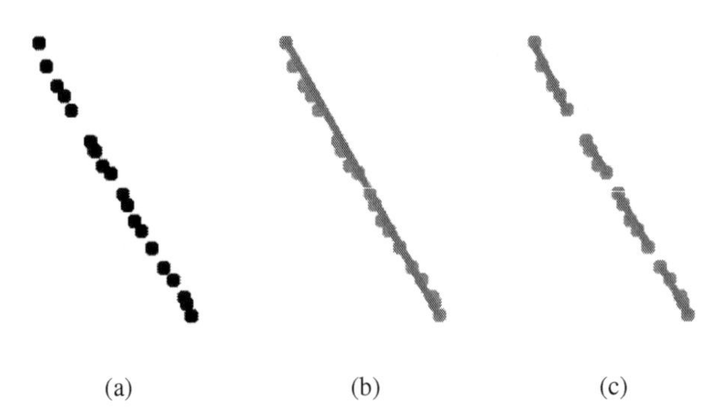

Fig. 6.16: Unreliable line detection: A wall perceived by a laser range finder is represented as points in 2-D (a). A line fitting algorithm fits a line through these points (b). With some probability, the fitting algorithm will erroneously interpret the data as several shorter lines (c)

Empirical results on the behavior of le-RLPR under distorted landmark detection are described in Sect. 7.3.2.

6.5.2 Robustness of Structure Space Representation

Because RLPR abstracts from detected line segments to a few binary features, the qualitative differences of varying observations are very high. The classification of line segments within the RLPR grid is of critical importance to represent the structure of the surrounding world.

In real-world scenarios, sensory input is affected by noise. In particular, the use of a line detection algorithm as a conceptual classification from laser range finder data to line segments is needed (see Example 4.4, p. 49). When using such a method, one cannot expect each wall to be classified as one coherent line segment by the line detection algorithm. In spite of that, it will frequently be detected as several disconnected shorter ones (see Fig. 6.16).

Fortunately, RLPR is very robust towards such errors: $\overline{\overline{\tau}}$ (used for the immediate surroundings) regards the existence of features in an area, because it does not distinguish between whether one or more than one line segment is within a region R. Besides, misclassifications are only to be expected when encountering borderline cases, for example, when the end of the line segment has not been detected correctly and so the overlap status for a region is wrong. But as these are borderline cases anyway, their impact on the overall expressiveness of the representation is hardly noticeable.

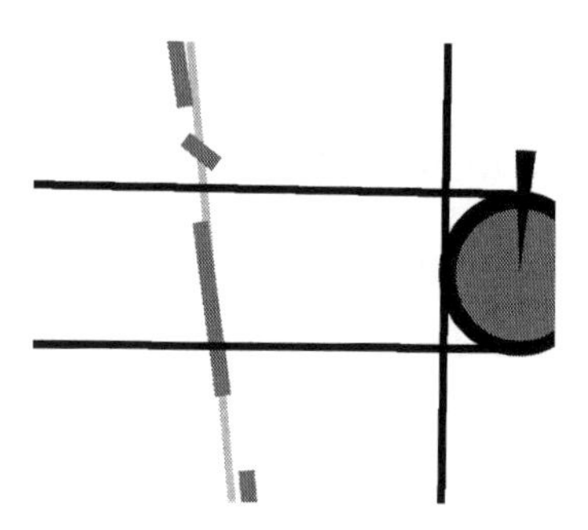

Fig. 6.17 Misclassifications in RLPR adjacency status: Only when the hole between line segments is exactly at one of the region boundaries are the RLPR values wrong

Misclassifications in the adjacency status $\overline{\overline{\tau}}'$ can have the severe effect that a whole wall disappears from the structural configuration of the current state. However, $\overline{\overline{\tau}}'$ delivers a wrong result only in cases where a "hole" between b_i and b_{i+1} is *exactly* at one of the region boundaries (see Fig. 6.17). In this case, no line segment crosses the region boundary and $\overline{\overline{\tau}}' = 0$. If the hole is not at a region boundary, one of the detected smaller line segments crosses this borderline; thus, $\overline{\overline{\tau}}' = 1$. Whether one long line segment or one of the shorter detected line segments crosses the boundary does not make a difference. Assuming we have a more or less meaningful line detection, such misclassifications will be the exception to the rule and will be observed rather rarely.

To cope with the remaining deficiencies in line detection, the following procedures can be used to minimize their effect:

- Usually, a line detection algorithm can be parametrized with regard to how likely two neighboring points are to belong to the same line segment, that is, whether the algorithm will detect line segments near each other as the same line segment or as two different line segments. If the likeliness to close a gap is chosen too high, gaps between line segments may not be noticed. For RLPR classifications, this is even a desirable effect: If the gap between two walls is too small for the robot to pass through, there is no benefit in considering it. So this parameter should be chosen high enough to merge close line segments and to ignore small gaps.
- To close gaps, one can also use *line segment prolongation*. This means that the system automatically prolongs each line segment after detection by a few centimeters at both ends. This also results in gaps being closed, regardless of whether those connected line segments emerged due to errors in the detection or really exist in the environment.

In the presented reinforcement learning framework (see Sect. 2.2), there are recurring observation cycles. If a serious error in one observation is encountered, it will most probably not persist in the next one, as long as we have non-systematic errors such as the ones described above. One can expect a wrong classification only to cause an erroneous reaction at one point in time, so the negative impact on the agent's behavior is limited. Actually, non-systematic errors are included within the transition probability T of the MDP model, so their effect has already been taken into account by the learned policy π.

An empirical evaluation of the robustness of le-RLPR under noisy line detection can be found in Sect. 7.3.2.2. Section 7.5 demonstrates the usage of RLPR in a real-world robot navigation scenario.

6.6 Summary

This chapter presented the landmark-enriched relative line position representation (le-RLPR) as a specialized formalism to support agent navigation in indoor environments. Based on the insights into abstraction and generalization derived in the previous chapters, le-RLPR is especially suitable when reinforcement learning is used.

le-RLPR is a structure space aspectualizable abstraction that consists of separate feature vectors for task and structure space. The task space can be described as a vector of landmarks detected relative to the agent's moving direction. To use landmarks only at decision points, structure-induced task space aspectualization SITSA has been introduced. For structure space, we defined the relative line position representation (RLPR) that abstracts sensory information about obstacles to a compact description of the structural surroundings of the moving agent. RLPR abstracts from unnecessary details and provides non-local generalization within a learning task.

The combined representation, landmark-enriched RLPR (le-RLPR) offers a discrete abstraction that shrinks the state space considerably. It is highly accessible because it is aspectualizable with regard to task space and structure space, and a high π^*-preservation can be achieved. A high amount of generalization can be reached, and le-RLPR is fully suitable for structure space transfer. Thus, le-RLPR allows reinforcement learning to be applied to the complex problem of indoor navigation in a large continuous state space. As both task and structure space descriptions also are very robust to sensor noise and wrong measurements, le-RLPR is suitable for application in real-world robotics.

Chapter 7
Empirical Evaluation

This chapter contains an empirical evaluation of the learning performance and generalization and transfer learning capabilities of RLPR-based representations in a goal-directed reinforcement learning task in an indoor office environment. It subsumes experiments performed over a longer period of research, so part of the results originate from peer-reviewed journal or conference contributions published by the author. Section 7.1 describes the experimental setup and the methods used for evaluation. In Sect. 7.2 we investigate the performance of le-RLPR in different variants. This section also includes experiments on task space tile coding (Sect. 7.2.4) and SITSA (Sect. 7.2.7). The robustness of le-RLPR under the influence of environmental distortion is investigated in Sect. 7.3, before the capabilities of le-RLPR regarding generalization and transfer learning are evaluated in Sect. 7.4. A case study on transferring knowledge learned in a simulator to a real-world environment is presented in Sect. 7.5. The chapter ends with a summary in Sect. 7.6.

7.1 Evaluation Setup

Before presenting different parts of the evaluation, let us first take a look at the testbed used for the learning experiments and the applied evaluation methods.

7.1.1 The Testbed

The testbed of all the experiments within this chapter is a basic simulated indoor environment for goal-directed robot navigation. It consists of a set of line segments arranged to form rooms and corridors. This testbed implements the scenario described in Sect. 6.1.1.

The simulated robot is a circular object with a diameter of 50 cm. It is capable of performing three different basic actions: moving forward and turning to the left and

L. Frommberger, *Qualitative Spatial Abstraction in Reinforcement Learning*,
Cognitive Technologies, DOI 10.1007/978-3-642-16590-0_7,

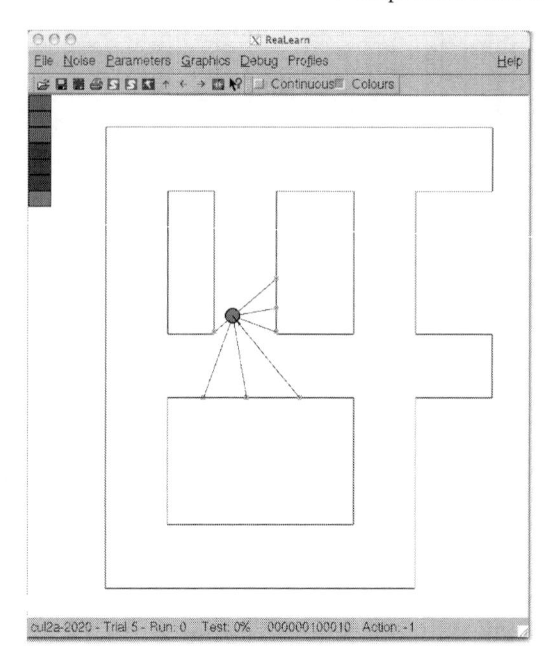

Fig. 7.1 Screenshot of the simulator program. Scans for wall color emit from the robot. The colors detected by the seven color scans are depicted in the upper left

to the right. The forward movement is set to 2.5 cm and the turns rotate the robot by approximately 4° (0.07 in radians) while additionally also moving the robot 2.5 cm forward. A configurable amount of noise is added to all actions; this noise model is described in the next section.

The robot is able to perceive walls and landmarks around it within a certain range. For all the experiments in this chapter, a range of 3.5 meters (seven times the robot's diameter) has been chosen. Objects or landmarks beyond this range are disregarded.

The walls within the environment can be uniquely identified by their color (sampled view model, see Sect. 6.2.4.1). Because of the benefits for evaluation described in Sect. 6.2.4.3, this model is used in all of the experiments except the one described in Sect. 7.2.6, where we regard point-based landmarks in the scene that are not physical objects, that is, they are not obstacles to the moving robot.

Figure 7.1 shows a screenshot of the simulator environment.

7.1.2 The Motion Noise Model

A noise model in the simulator affects every action of the agent with a configurable amount of distortion. An action consists of a forward movement $\Delta x \in \mathbb{R}^+$ and a turn $\Delta\theta \in [-\pi, \pi]$. At each time step a distortion vector (N_0, N_1) is calculated and added to the motion vector $(\Delta x, \Delta\theta)$ such that we achieve a "real" movement $(\Delta x + \zeta N_0, \Delta\theta + \zeta N_1)$. (N_0, N_1) follows a Gaussian distribution, and the factor $\zeta \in \mathbb{R}_0^+$ determines the strength of the distortion.

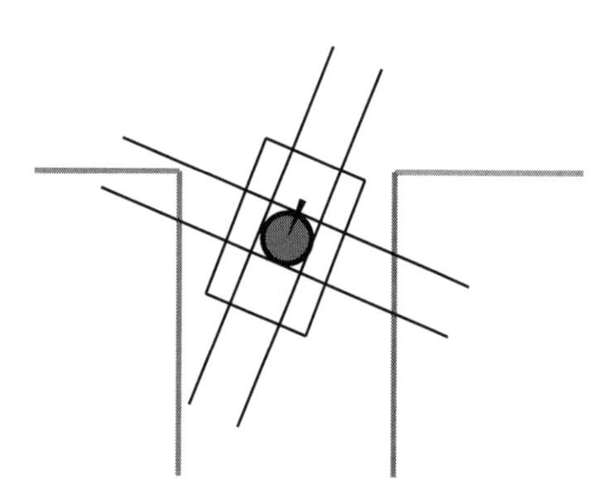

Fig. 7.2 The RLPR grid drawn around the robot. The immediate surroundings extend 50 cm to the left and right and 1 m to the front and back of the robot

7.1.3 The le-RLPR Representation

For all the experiments with colored walls the robot scanned the environment at discrete angles between -90° and 90°. If not otherwise stated, seven color scans have been used.

In all the experiments with le-RLPR, the size of the immediate surroundings of the RLPR grid has been set to 1×1.5 meters, that is, the inner region extends 25 cm to the left and right and 50 cm to the front and back of the robot (see Fig. 7.2).

7.1.4 Learning Algorithm, Rewards, and Cross-validation

To evaluate the use of the presented spatial representation, we keep the setup of the learning experiment as simple as possible. For all the experiments standard methods are used and complex tricks or smart heuristics are avoided. Even if they would increase the learning performance, we would move the focus away from the use of spatial abstraction. Furthermore, one aim of this work is to show that a well-considered choice of representation can be effective and successful even without applying expert knowledge to set up reinforcement learning settings.

The off-policy method Q-learning (see Sect. 2.5.3) is especially attractive because it does not require a model. Thus, all the experiments have been conducted using this learning algorithm. In particular, due to the beneficial effect on perceptual aliasing (Sect. 6.2.5.1), we use replacing eligibility-traces (Sect. 2.5.2) with Watkins's Q(λ) algorithm (Watkins, 1989).

For exploration we stick to the most frequently used ε-greedy exploration strategy (see Sect. 2.4.1) with an exploration rate of $\varepsilon = 0.15$ if not stated otherwise. The reward function R is designed such that the agent receives a positive reward of 100 for reaching the goal state and a negative one of -100 for colliding with a wall.

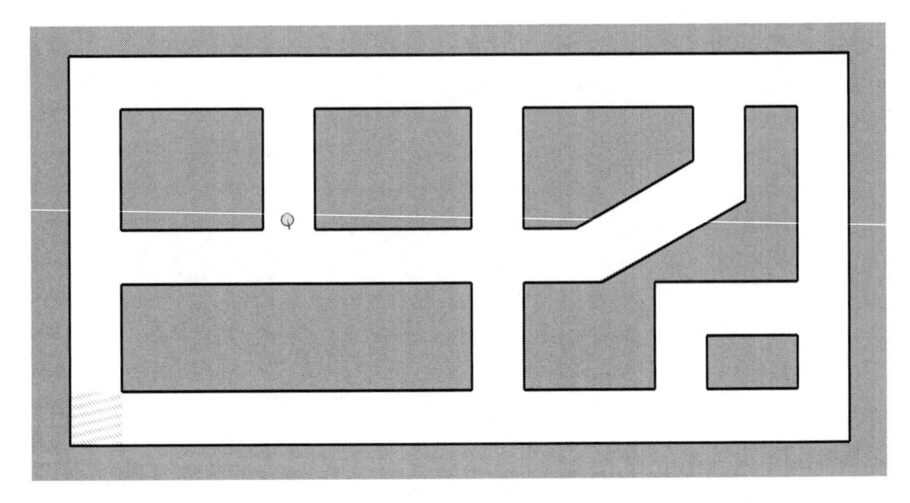

Fig. 7.3: The simulated indoor robot environment used for evaluation. The goal area is marked in the lower left corner

In all other states, no reinforcement signal is given. The discounted reward model (see Sect. 2.3.2) was used in all the trials.

The parameters (learning rate α, exploration rate ε, discount γ, and so on) may vary from experiment to experiment. Because parameter fiddling is not a goal of this evaluation, it is not claimed that these parameters lead to the best possible performance. Because of the fact that experiments have been conducted over a longer period of time and originate from various test periods prior to the writing of this book, parameter choices might differ without an obvious reason. However, in no case are these differences the result of needed adaptations to prove the success of any method.

In each learning episode, the robot starts from one predefined starting position within the environment. In most of the experiments 1,000 starting positions have been randomly set within the corridors. While the majority of them have been used for training, a subset (usually 200) of the starting positions has been used for cross-validation: After a certain number of training episodes, test runs without learning are started from this cross-validation set to check the success of the policy up to that moment. The cross-validation set of starting positions is disjoint from the one used for training.

7.2 Learning Performance

In this section we take a look at the performance of learning goal-directed navigation when using le-RLPR. All the experiments described in Sect. 7.2.1 through Sect. 7.2.6 took place in the environment depicted in Fig. 7.3.

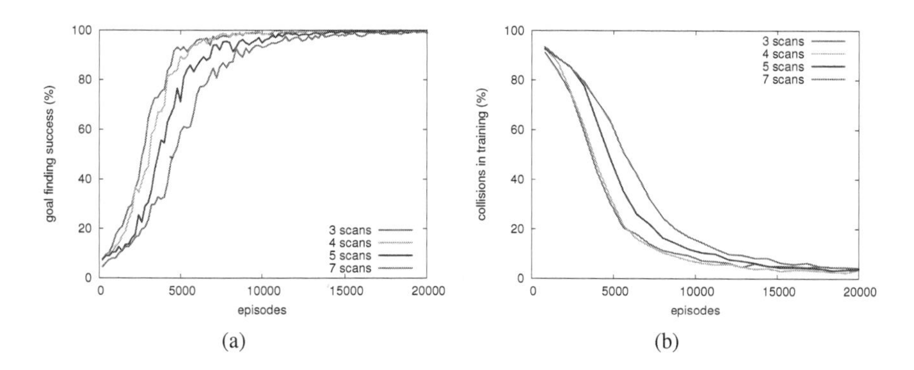

Fig. 7.4: Comparison of le-RLPR with 3, 4, 5, and 7 landmark scans. For both goal finding success (a) and the number of collisions in training (b), all variants show comparably good performance

7.2.1 Performance of le-RLPR-Based Representations

The first experiment investigated the influence of two important parameters of le-RLPR, the number of landmark scans and the design of the RLPR grid. Throughout this section a learning rate $\alpha = 0.01$, a decay rate $\gamma = 0.98$, and $\lambda = 0.9$ have been used.

7.2.1.1 Number of Landmark Scans

To analyze the influence of the number of landmark scans on the learning performance, the agent was trained with RLPR_1 (see Fig. 6.13) as the RLPR grid layout using 3, 4, 5, and 7 scans ranging from $-90°$ to $90°$ with regard to the agent's moving direction. The goal finding success for every number of scans is shown in Fig. 7.4a. A larger number of scans slows down the learning speed due to the larger state space size, but the difference is not too large: All configurations reach a near-100% goal finding success before 15,000 learning episodes.

The number of collisions that occur in training over time is plotted in Fig. 7.4b. The lower the number of landmark scans, the fewer the collisions, with the exception that three scans produce slightly more collisions than four scans. Also, the learning speed with three scans is only very moderately higher than that with four scans, so it can be concluded that for this kind of problem four scans is the optimal number. Note that the number of collisions never becomes 0 because of the random exploration during training.

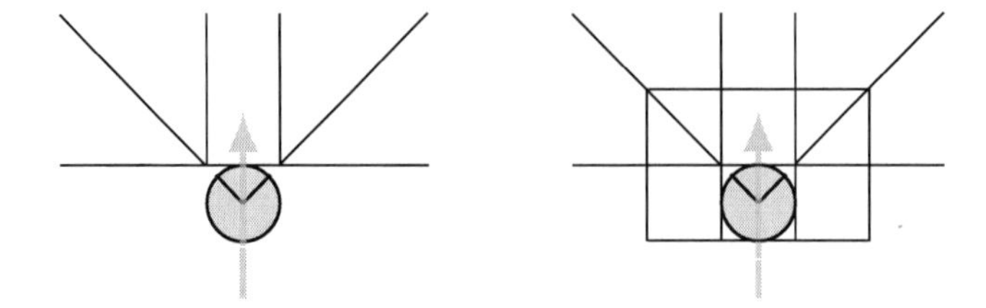

Fig. 7.5: $RLPR_5$ is a reduced version of $RLPR_1$ that is tailored to the motion dynamics of the robot in the given scenario. Information behind the robot is completely omitted

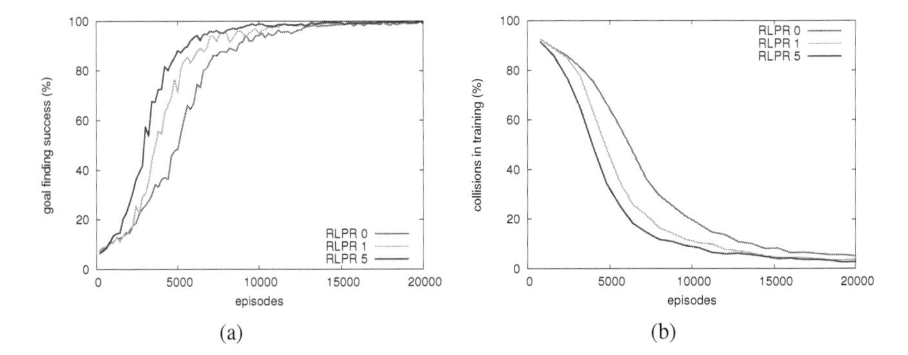

Fig. 7.6: Comparison of the different RLPR grid variants $RLPR_0$, $RLPR_1$, and $RLPR_5$. Even if creating a larger state space, $RLPR_1$ learns faster (a) and produces fewer collisions (b) compared to $RLPR_0$. $RLPR_5$ again slightly improves the performance

7.2.1.2 RLPR Grid Variants

The next experiment tested the RLPR grid variants $RLPR_0$ and $RLPR_1$ as described in Sect. 6.3.3 as well as another RLPR variant called $RLPR_5$ that is depicted in Fig. 7.5. $RLPR_5$ is derived from $RLPR_1$ by omitting all information at the back of the robot. This is tailored to the current scenario where the robot is incapable of backward movement and thus information about the environment behind it does not play a critical role.

The success rates for le-RLPR with five landmark scans with different grid variants are shown in Fig. 7.6a and the development of training collisions is shown in Fig. 7.6b. Even if creating a larger state space due to two additional features, $RLPR_1$ enables faster learning and leads to considerably fewer collisions compared to $RLPR_0$. This is an indicator for the appropriateness of $RLPR_1$ for the given navigation task (compare to Fig. 6.14). The reduced variant $RLPR_5$ again slightly improves the learning performance due to the smaller observation space size.

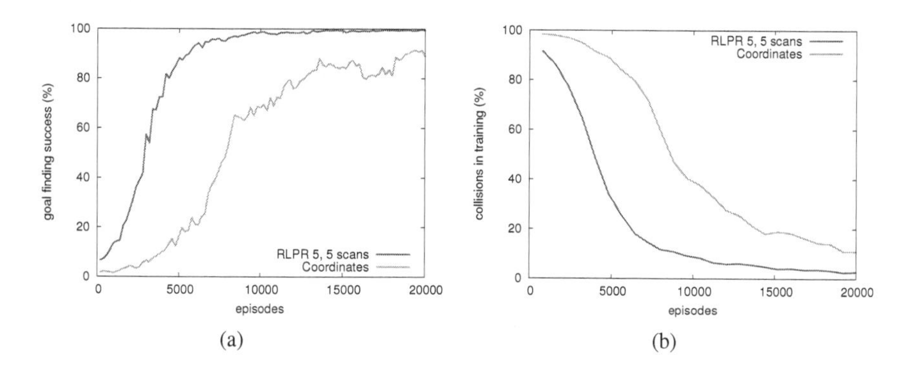

Fig. 7.7: Performance of le-RLPR in comparison with a representation of the original MDP

7.2.2 le-RLPR Compared to the Original MDP

The next experiment compares the performance of an RLPR-based representation with the results when using the original MDP representation, that is, with a state s being represented as the robot's coordinates and orientation: $s = (x, y, \theta)$.

To achieve learning success in this continuous original MDP state space, CMACs (see Sect. 5.3.2) have been used for value function approximation. Both the "ordinary" CMAC and the A-CMAC have been investigated.

Various combinations of coarsenings on coordinates and angle have been tried, starting from coordinate grids of $10\,\mathrm{cm} \times 10\,\mathrm{cm}$ and an angular resolution of $5°$ to $1\,\mathrm{m} \times 1\,\mathrm{m}$ and $20°$. The best performance could be achieved with a grid of $50\,\mathrm{cm} \times 50\,\mathrm{cm}$ and a resolution of $10°$ with an ordinary CMAC with ten tilings and the parameters $\alpha = 0.02$, $\gamma = 0.98$, and $\lambda = 0.9$. Figure 7.7 shows success rate and collisions of this setting compared to le-RLPR with five scans and RLPR_5 as described above.

With le-RLPR, the goal finding success reaches 100% after about 10,000 episodes. At the same time, the best coordinate-based representation could only achieve a success rate of 75%, and the system failed in continuously reaching more than 95% success rate within 100,000 episodes. With different parameters (for example, with a $25\,\mathrm{cm} \times 25\,\mathrm{cm}$ grid and $15°$ angular resolution), near-100% learning success could be reached after about 40,000 episodes, but with a much slower learning in the early phase. For all the coordinate-based experiments it holds that learning was either very slow compared to le-RLPR or affected by a suboptimal or unstable goal finding success.

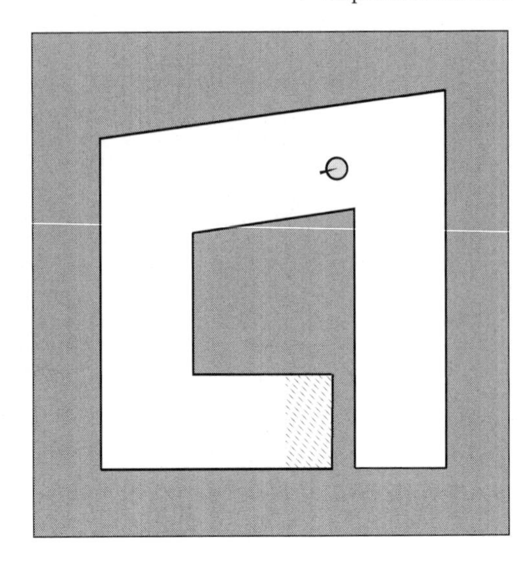

Fig. 7.8 A tiny environment without decision points

7.2.3 Quality of le-RLPR-Based Solutions

In the next experiment it has been analyzed how the use of le-RLPR affects the *quality* of the solution. Because abstraction lowers the resolution of the state space and leads to reduced exactness it can be expected that the learned policy is not optimal with regard to the achieved reward. In this scenario this means that when following policies learned with le-RLPR the agent follows a longer trajectory from starting position to the goal state than needed.

Obtaining the optimal solution in the given scenario is not trivial, because we have a stochastic process and the effects of motion noise cannot be predicted. Thus, finding an optimal policy is a perfect use case for Q-learning, which is known to converge to an optimal solution (see Sect. 2.5.3). However, in a continuous state space, value function approximation is mandatory. Thus, an approximation close to the states of the original MDP was used. In detail, coordinates were approximated by grid cells of $0.5\,\text{cm} \times 0.5\,\text{cm}$ and the angular resolution of the orientation was set to 1°. These values proved to be a reasonable approximation that still allowed for solving the problem in manageable time using a CMAC with ten tilings for value function approximation.

For the environment in Fig. 7.3 the learned solution did not show convergence to an optimal solution even after 5,000,000 episodes of training. Thus, this experiment has been conducted in a very small environment without any decision points, depicted in Fig. 7.8. In this world the agent was trained with the approximate representation given above for 5,000,000 episodes with $\alpha = 0.01$, $\gamma = 0.98$, and $\lambda = 0.9$ such that overall reward and, thus, travel distance converged to a stable value. The distance was measured from 100 starting positions in the cross-validation set. Even in this tiny world, each trial took more than a week to compute.

Table 7.1: Traveled distance of the robot under different representations: Compared to the result with a representation that closely approximates the original MDP, le-RLPR leads to slightly longer trajectories. Especially $RLPR_1$ performs almost as good as the coordinate representation. The larger the noise level, the smaller the loss.

	original MDP	$RLPR_0$		$RLPR_1$	
noise	travel	travel	increase	travel	increase
0.7	172132	184306	+7.1%	176443	+2.5%
1.8	17813	186451	+4.6%	180816	+1.5%

This distance has been compared to the results when applying $RLPR_0$ and $RLPR_1$ for 1,500,000 episodes ($\alpha = 0.01$, $\gamma = 0.98$, $\lambda = 0.9$). Table 7.1 summarizes the results for two motion noise levels, $\sigma = 0.7$ and $\sigma = 1.8$. With $\sigma = 0.7$, the distance the robot traveled is 7.1% longer with $RLPR_0$ and 2.5% longer with $RLPR_1$. For the noisier setup ($\sigma = 1.8$), the increase was only 4.6% ($RLPR_0$) and 1.5% ($RLPR_1$). So, especially with $RLPR_1$, the quality of the gained solution did not drop significantly. The smaller quality loss for a higher level of motion noise is a sign of good robustness of le-RLPR against distortion. This will be investigated in detail in Sect. 7.3.

7.2.4 Effect of Task Space Tile Coding

Task space tile coding (TSTC) was introduced in Sect. 5.3.3 as a method to enable non-local generalization in structure space aspectualizable state spaces based on structural knowledge within the same task. The next experiment evaluated the effect of TSTC. le-RLPR with seven landmark scans and $RLPR_0$ was used in the scenario in Fig. 7.3, with and without TSTC, with $\gamma = 0.98$ and $\lambda = 0.9$. For the non-TSTC task, a learning rate of $\alpha = 0.02$ was chosen while $\alpha = 0.005$ was used for the TSTC case. Due to the strong generalization of TSTC which leads to an increased number of value function updates per step, lower learning rates are mandatory. The results of the experiment are plotted in Fig. 7.9.

The effect of task space tile coding is dramatic. After a few dozen episodes it becomes visible that the agent develops a way to cope with structures in the world and begins to follow corridors and turn around curves. This leads to a thorough exploration of the environment very quickly. Thus, near 100% goal finding success is reached after 1,000 episodes instead of 10,000 episodes without TSTC. The decrease in collisions is even stronger: After 15,000 episodes of learning the non-TSTC agent still collides more often than the TSTC one after 1,000 episodes.

Experiments on the effect of task space tile coding under distorted perception follow in Sect. 7.3.2

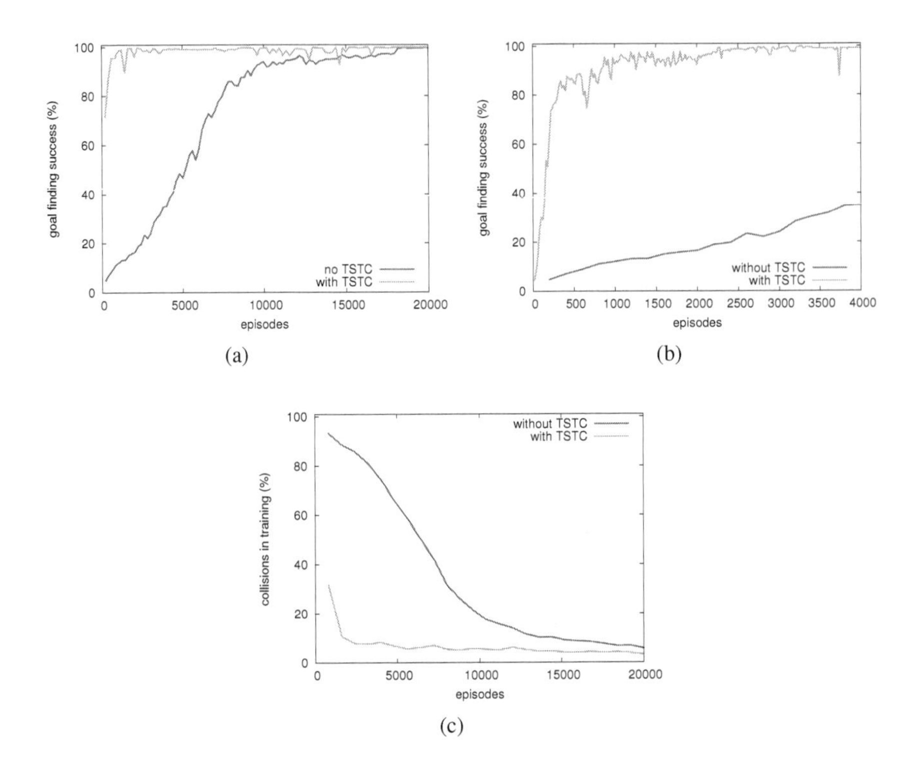

Fig. 7.9: Effect of task space tile coding: Compared to "ordinary" le-RLPR, the goal finding success increases rapidly with TSTC (a). In (b), the comparison is shown on a smaller scale (the results are not originating from the same trial as (a)). The number of collisions in training soon becomes negligible with TSTC (c)

7.2.5 Task Space Information Only

Section 6.2.5 discussed how navigation can be realized just by relying on the information on landmarks, and the problem of perceptual aliasing that arises then. However, with the extended landmark model established by the uniquely colored walls, rich landmark information can be retrieved at any position within the environment. With a large number of landmark scans, this might suffice to minimize perceptual aliasing and allow for learning a successful navigation behavior without additional use of RLPR. Therefore, the performance using $\psi_T(s)$ as state representation with 7, 10, and 15 color scans ($\alpha = 0.002$, $\gamma = 0.98$, $\lambda = 0.9$) was investigated.

The result is depicted in Fig. 7.10. With seven scans, no 100% goal finding could be achieved. With ten and 15 scans, learning was slightly unstable, but successful after 25,000 or 50,000 episodes (Fig. 7.10a). With ten scans, the robot learned even faster than with seven scans. This result is confirmed when looking at the number

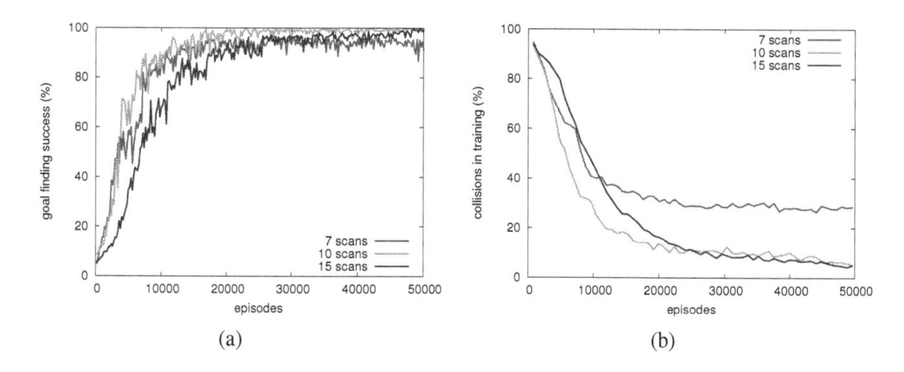

Fig. 7.10: Learning with landmark information only. From ten scans on, a nearly stable goal finding success could be achieved (a), while seven scans do not suffice and produce many collisions in training (b).

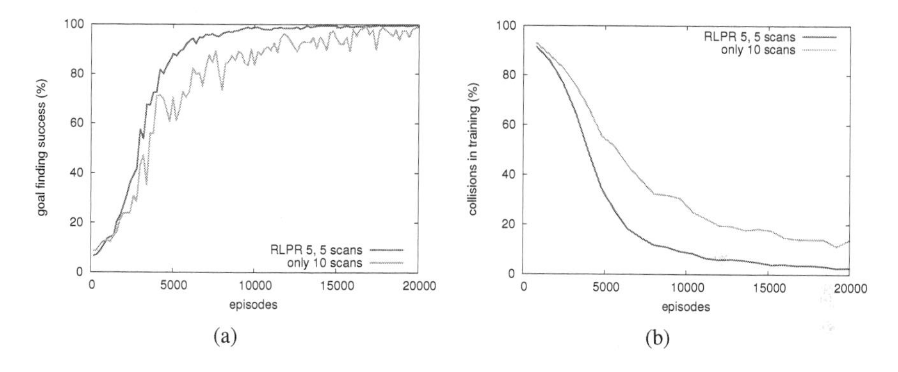

Fig. 7.11: Ten landmarks scans only compared to le-RLPR with five scans and $RLPR_5$: After a similar early training phase, le-RLPR enables robust 100% goal finding success much earlier (a). Without RLPR, the agent collides frequently in training (b)

of collisions in training (Fig. 7.10b): With seven scans, more than 35% of the trials ended in a collision even after 50,000 episodes.

A large number of landmark samples can be used successfully for representing the state space. However, compared to le-RLPR, landmark-only representations are the worse alternative. Figure 7.11 shows the performance of ten colors scans only compared to $RLPR_5$ and five color scans as used in Sect. 7.2.2. With respect to both goal finding success and training collisions, le-RLPR performs much better, especially with respect to robustness and the point where 100% goal finding success is reached. Furthermore, landmark-only representations cannot benefit from structure-based generalization methods such as task space tile coding or structure-induced task space aspectualization (SITSA, see Sect. 5.6).

7.2.6 Learning Navigation with Point-Based Landmarks

The use of the sampled view landmark model with colored walls has been favored for evaluation because the number and location of point-based landmarks within the world has critical influence on the learning speed (see Sect. 6.2.4.3). This section shows this influence on the learning performance. Furthermore, it is demonstrated that le-RLPR also works fine with point-based landmarks.[1]

Figure 7.12 shows a world with different distributions of landmarks, ranging from a small to a very large number of landmarks, and also a world with landmarks only at intersection points.

For representing landmark information in these worlds the landmark mapping procedure described in Sect. 6.2.3 was used with eight sectors ranging from $-140°$ to $140°$ around the agent without landmark selection (Sect. 6.2.3.1). The learning parameters were a step size of $\alpha = 0.2$, a discount factor of $\gamma = 0.98$, and $\lambda = 0.9$.

Success graphs of this experiment are shown in Fig. 7.13a. Goal finding success critically depends on the number of landmarks—the more there are, the longer are the observed learning times. With few landmarks the learning performance is extremely good. Furthermore, it gets obvious that landmarks only at intersections are sufficient to succeed in the task. This shows how a clever choice of landmark distribution over the world can influence the learning behavior, which makes comparisons for evaluation difficult.

For the huge number of landmarks in the world in Fig. 7.12d, a stable success rate of 100% was not reached even after 50,000 learning episodes. Thus, the landmark selection strategy based on distance as described in Sect. 6.2.3.1 was applied with a maximum number of $l_{\max} = 1, 2$, or 3 landmarks per sector. The result is shown in Fig. 7.13b: While the "all landmarks" representation does not reach a stable success rate of 100% after 50,000 learning episodes, selection by "close landmark first" manages to do so after about 25,000 ($l_{\max} = 1$) or 38,000 episodes ($l_{\max} = 2$). For $l_{\max} = 3$ the strategy shows fast learning in the early phase, but converges slower towards 100% success. So one landmark per sector is enough to fulfill the task in the given scenario.

[1] The results in this section originate from experiments published in Frommberger (2008b).

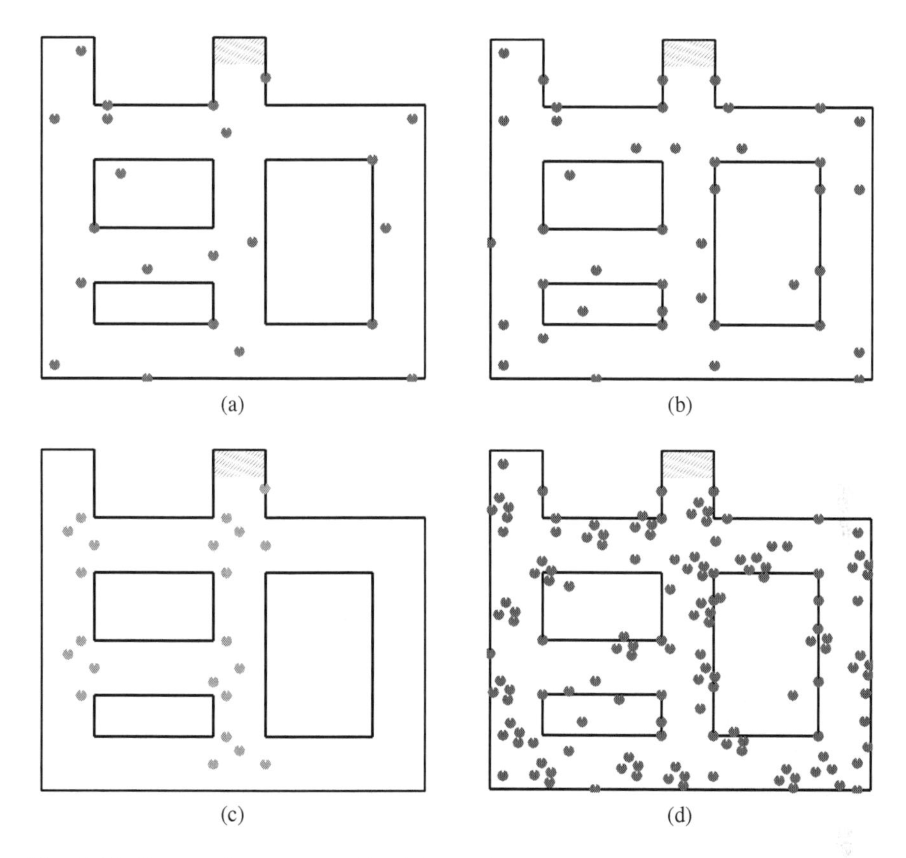

Fig. 7.12: Four environments with different distributions of landmarks, represented by dots: few landmarks (a), many landmarks (b), landmarks placed only at intersections (c), and a very high number of landmarks distributed all over the place (d)

7.2.7 Evaluation of SITSA

This section evaluates structure-induced task space aspectualization (SITSA), as described in Sect. 5.6, applied to the landmark part of le-RLPR (SDALS), as suggested in Sect. 6.2.3.3.[2]

For determining non-decision structures for SITSA, a successful policy for a goal finding task in the world depicted in Fig. 7.1 was analyzed using the procedure described in 5.6.2 and a set $\mathcal{S}_{\mathrm{NDesc}}$ of 110 non-decision structures was derived from it. The evaluated learning task then took place in the environment depicted in Fig. 7.3. For representing $\perp$, the vector (0,0,0,0,0,0,0) was used. This landmark representation is impossible to perceive, because we start to enumerate wall numbers with 1.

[2] Some of the results in this section originate from Frommberger (2009).

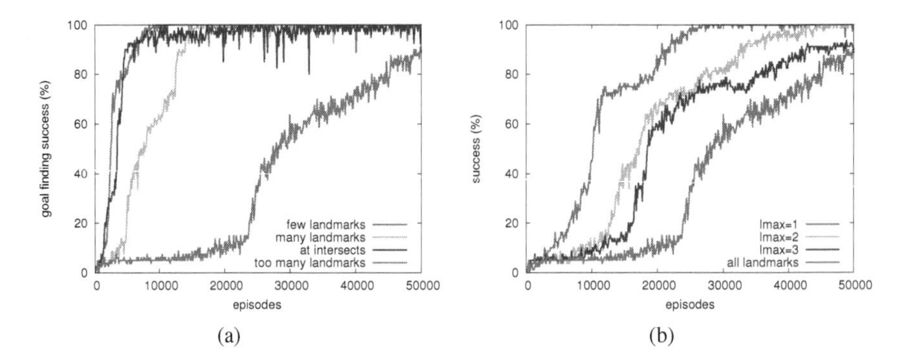

Fig. 7.13: Goal finding success in the environments with a different number of landmarks (Fig. 7.12): Few or appropriately placed landmarks enable faster learning, while too many landmarks hinder the learning progress (a). Landmark selection by distance resolves the problem of too many landmarks (b): When selecting one or two landmarks per sector, 100% goal finding success can be reached before 50,000 learning episodes

In the new learning task the agent started from 1,900 starting positions randomly distributed over the corridors. Every 500 episodes, the success was evaluated on a cross-validation set of 100 starting positions disjoint from the ones used for training. A step size $\alpha = 0.01$, a discount factor $\gamma = 0.99$, $\lambda = 0.9$, and an exploration rate $\varepsilon = 0.2$ were used in training with le-RLPR with five color scans and RLPR_1.

Figure 7.14 shows the performance of the agent with and without the use of SITSA. When using the situation-dependent abstraction SITSA offers, the learning speed is improved, especially in the early learning phase. With SITSA, a stable 95% goal finding success is reached after 6,500 episodes while it takes 15,000 episodes without it. However, this comes at the price of an increased number of collisions during training.

One thing that became obvious over the experiments was that the number of starting positions is critical for the use of this abstraction: When using very few starting positions (for example, 20 instead of 1,900), the learning system occasionally ran into the exploration-exploitation dilemma: Because the agent generalizes very fast it acquires a generally sensible behavior early. In the following, the agent mainly follows this behavior while learning, which leads to insufficient exploration. A more sophisticated exploration method than ε-greedy policies might be needed here.

7.3 Behavior Under Noise

Qualitative representations have the property of sharp boundaries between concepts, that is, something is represented or not—there is nothing "fuzzy" in them. So quali-

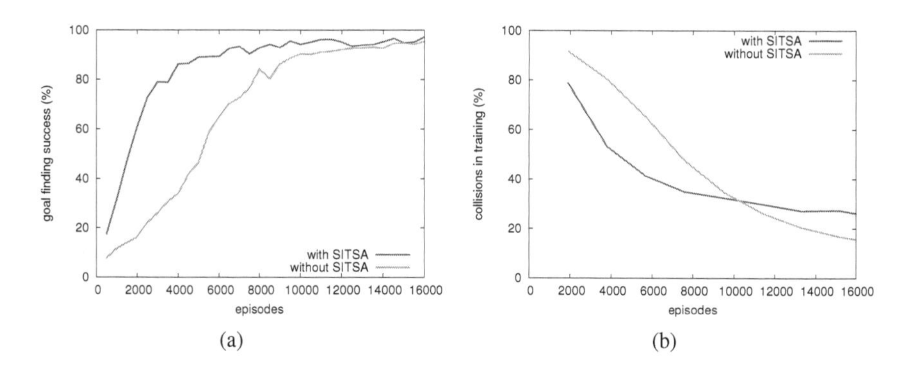

Fig. 7.14: Comparison: learning success with SITSA and without. With SITSA, learning is considerably faster (a) but leads to more collisions in training (b)

tative representations raise the question of whether this sharpness is appropriate for control tasks, especially when there is a noisy and uncertain environment to cope with. In this section we will investigate the properties of le-RLPR under distortion in both the agent's movement and its perception.

7.3.1 Robustness Under Motion Noise

First let us take a look at distortion in the agent's movement. le-RLPR was tested with five scans and RLPR_0 in the world in Fig. 7.3 with a noise parameter $\zeta \in \{0.7, 1.8, 2.5\}$ (see Sect. 7.1.2 for the motion noise model), and compared to a noise-free setting. The chosen parameters were $\alpha = 0.01$, $\gamma = 0.98$, and $\lambda = 0.9$. The result is plotted in Fig. 7.15: Lower noise levels have hardly any impact on the learning speed and the number of collisions in training. Only $\zeta = 2.5$ shows a noticeable loss of performance. The trial with $\zeta = 0.7$ even performed slightly better than the noise-free trial: We even benefit from a low amount of noise: The agent learns a less risky behavior.

In comparison, Fig. 7.16 plots the effect of the same set of distortion parameters on the coordinate-based representation as used in Sect. 7.2.2. The influence of motion noise is much stronger there compared to le-RLPR. While the latter almost did not change under noise levels of $\zeta \leq 1.8$, the coordinate-based representation suffers a lot from it. Summed up, le-RLPR shows very good robustness against distorted motion.

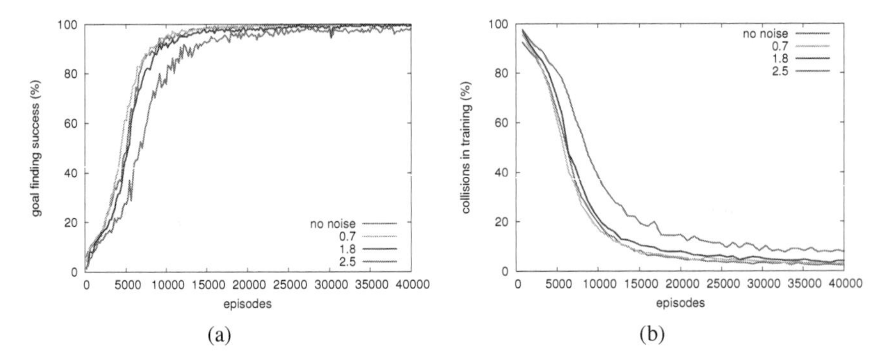

Fig. 7.15: Goal finding success under different levels of motion noise. Lower noise levels have hardly any impact on learning speed (a) and the number of collisions in training (b)

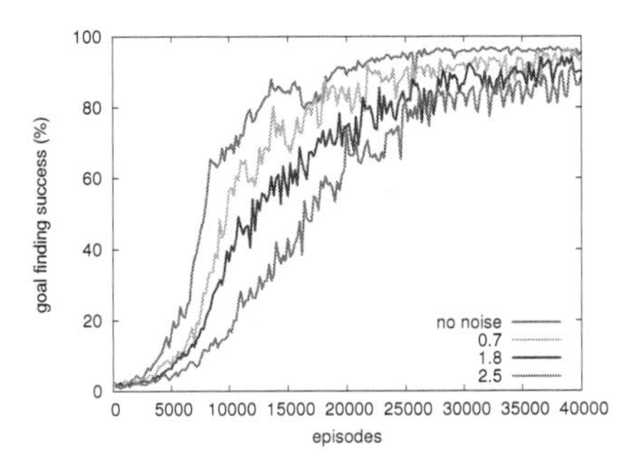

Fig. 7.16: Goal finding success under different levels of motion noise: the coordinate-based representation of the original MDP is heavily influenced by distorted motion

7.3.2 Robustness Under Distorted Perception

This section evaluates the robustness of le-RLPR towards environmental noise.[3] The theoretical aspects of robustness of le-RLPR have been laid out in Sect. 6.5.

7.3.2.1 Robustness Under Distorted Landmark Detection

First, it is tested how robust the representation is against distortion in landmark detection and how structural information of RLPR helps the system to cope with very unreliable landmark detection. Therefore, RLPR_0 with seven scans with the

[3] The experiments on distorted landmark detection described in this section have been carried out in Frommberger (2007a).

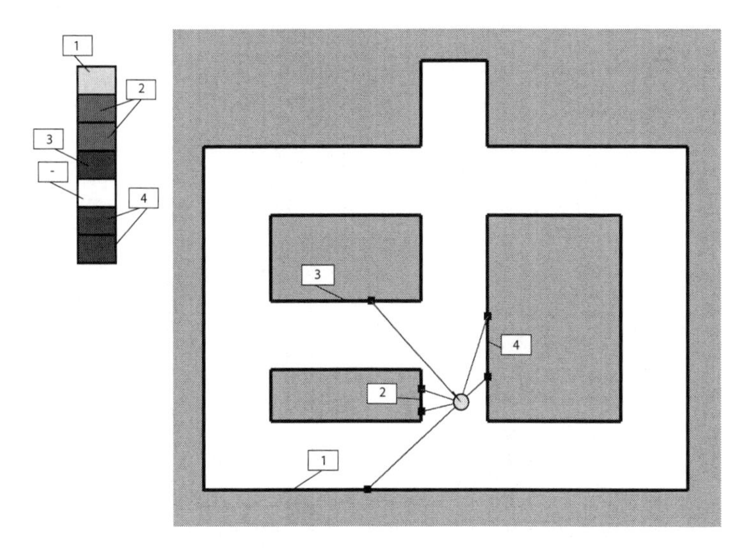

Fig. 7.17: A slightly modified office environment with only one dead end. Coloring of walls is not shown in this picture

application of task space tile coding (Sect. 5.3.3) was regarded. The parameters were the same as in the previous section. The experiment took place in the world depicted in Fig. 7.17.

The noise model implied here is the following: Each single landmark scan c_i returns the correct landmark detection with a probability of σ, so the probability that $(c_1,\ldots,c_7)$ is detected correctly is σ^7. For $\sigma = 95\%$ the overall probability of detecting the whole vector correctly is around 70%, 3% for $\sigma = 60\%$ and only 0.2% for $\sigma = 40\%$. In the case of a failure, no landmark is detected.

Figure 7.18 shows the learning success for values of $\sigma = 95\%$, 85%, 60%, 40% and 20%. The higher the failure in landmark detection, the lower the rate at which the agent reaches the goal. However, even with $\sigma = 60\%$, the goal is reached in about two thirds of the test runs after 40,000 episodes of training. This is still a very good value, because the goal area is located behind a crossing, and reaching it is subject to a decision, which is hard to make when the agent is that uncertain of where it is. Furthermore, the number of collisions is more or less negligible after very few runs, almost independently of σ. This shows that TSTC performs well in enabling the agent to learn a good general navigation behavior even in the absence of usable landmark information.

7.3.2.2 Robustness Under Distorted Line Detection

In Sect. 6.5 we have seen that misclassifications in the RLPR representation can only occur at region boundaries and thus RLPR shows high robustness with few misclassifications of minor impact. This is evaluated in this section.

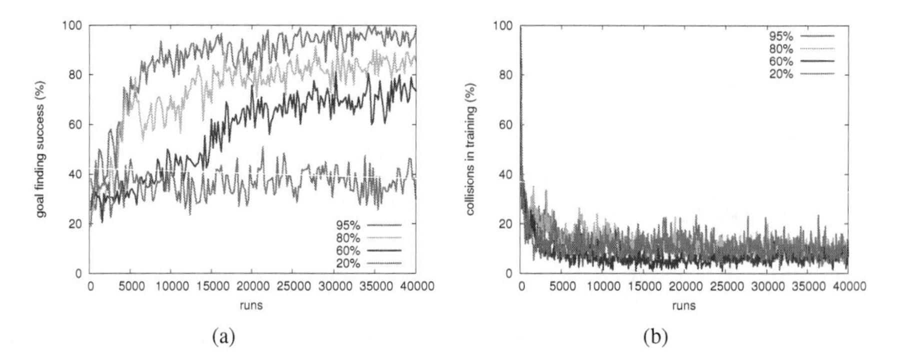

Fig. 7.18: Learning with unreliable landmark information for different levels of input distortion σ: The learning success of reaching the goal (a) decreases with higher values of σ, because the agent is uncertain of where it is; but it is still comparably high even for very high distortion levels. The number of collisions (b) is hardly affected by landmark distortion

The Noise Model

Another noise model has been applied to the line detection algorithm of the simulator program. Depending on a parameter $\rho \in \mathbb{N}$, $k \in \{1,\ldots,\rho+1\}$ holes of 20 cm each are inserted into each line segment l of a length of five meters; that means, a wall with a length of five meters will on average be detected as $\frac{1}{2}(\rho+1)+1$ line segments (for comparison, the robot's diameter is 0.5 meters). Additionally, the start and end points of the resulting line segments $l_1,\ldots,l_k$ are then relocated in a random direction. The result is a rather realistic approximation of line detection in noisy laser range finder data (see Fig. 7.19).

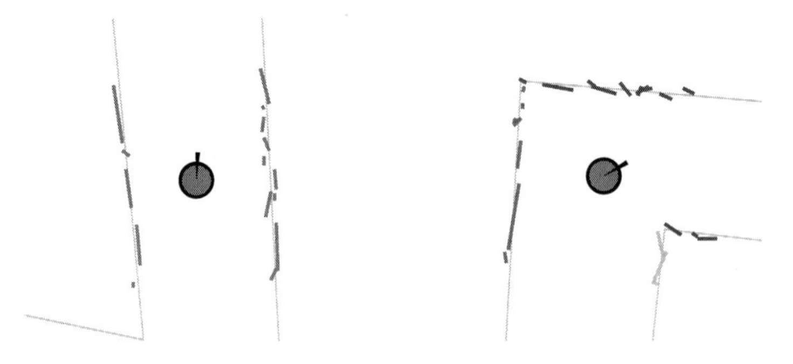

Fig. 7.19: The noise model: The real walls (thin lines) are detected as several small chunks (thick lines). Depicted is a noise level of $\rho = 20$

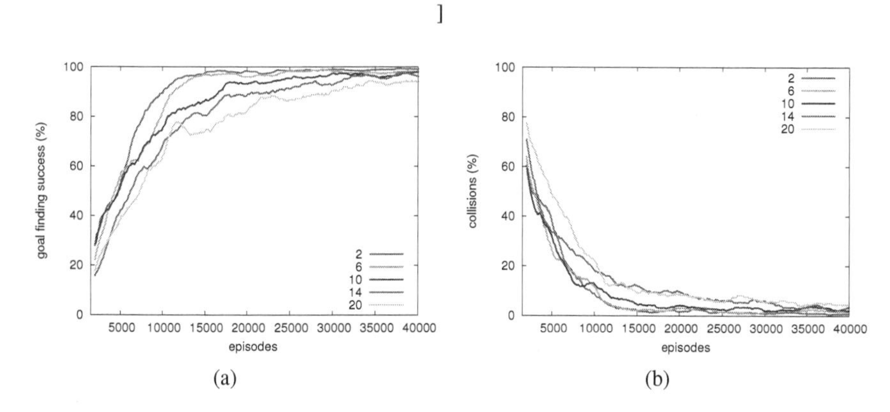

Fig. 7.20: Learning with unreliable line detection: Learning success (a) and collisions (b) with different noise levels ρ. Even under heavy noise, the system learns successfully (but somewhat slower) with a low number of collisions

Performance Under Noisy Line Detection

Figure 7.20a shows the manner in which the learning success is affected by different noise values ρ. Of course, as the noisy perception influences the transition probabilities T, learning time increases with the value of ρ and learning success decreases marginally. But up to $\rho = 20$, a success rate of 95% in tests can be reached after 40,000 learning episodes.

With respect to collisions, the differences between the noise levels are even smaller compared to the differences with regard to learning speed. Even with $\rho = 20$ (which means more than 11 detected line segments per five meters on average), the agent performs comparably well and does not collide often (Fig. 7.20b).

Figure 7.21 shows the effect of the line segment prolongation technique suggested in Sect. 6.5.2. Under a noise level of $\rho = 20$ this method leads to a significant improvement over the result achieved without prolongation: the learning performance is comparable to the behavior with very low noise levels ($\rho \leq 6$)).

7.4 Generalization and Transfer Learning

In this section we look at the manner in which le-RLPR can support generalization and knowledge transfer.

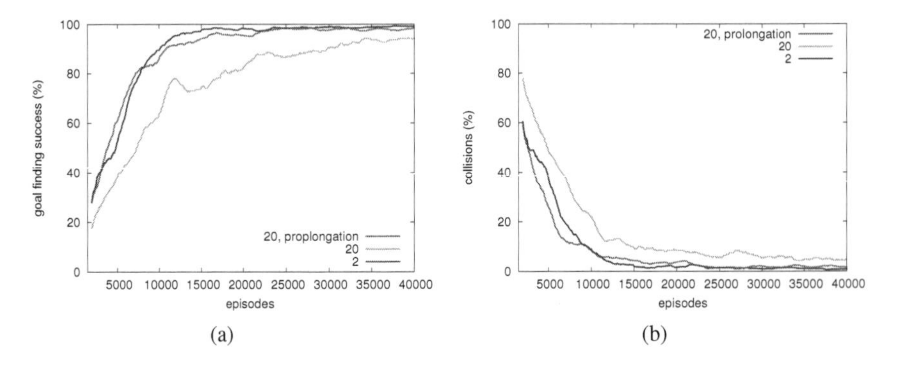

Fig. 7.21: Line segment prolongation significantly improves the learning behavior at noise level $\rho = 20$. With segment prolongation, the learning speed (a) and collisions in training (b) are comparable to the learning success with $\rho = 2$

7.4.1 le-RLPR and Modified Environments

le-RLPR is agent-centered and abstracts from metrical details. Thus, learned strategies are expected also to be successful in environments that differ from the original one with respect to dimension, orientation, and angles of corridors. To evaluate this, we take a look at the agent's behavior in modified environments that have the same topology (including the same coloring of walls) as the one that the agent has been learning in.[4]

Training has been performed for 40,000 episodes in the environment depicted in Fig. 7.17 using RLPR_0 and seven color scans, ($\alpha = 0.2$, $\gamma = 0.98$, $\lambda = 0.9$). The resulting policies have then been applied to two similar "deformed" environments: one where all the distances on the x-axis have been scaled down (Fig. 7.22a) and one with changed angles between the corridors (Fig. 7.22b).

The goal finding success of policies learned both with and without task space tile coding (TSTC) was tested in the deformed worlds. The agent started from 50 starting positions spread over the corridors. The results are given in Table 7.2: With and without TSTC, the policies show a successful goal finding performance. While without TSTC up to 12% of the test runs end in a collision, the agent found the goal in at least 98% of the cases with a TSTC-based policy. Trajectories produced by the agent following this policy are depicted in Fig. 7.22: The agent is able to show a generally sensible behavior, as described in Example 5.2 (p. 70), in both the worlds.

[4] This section describes an experiment originating from Frommberger (2008a).

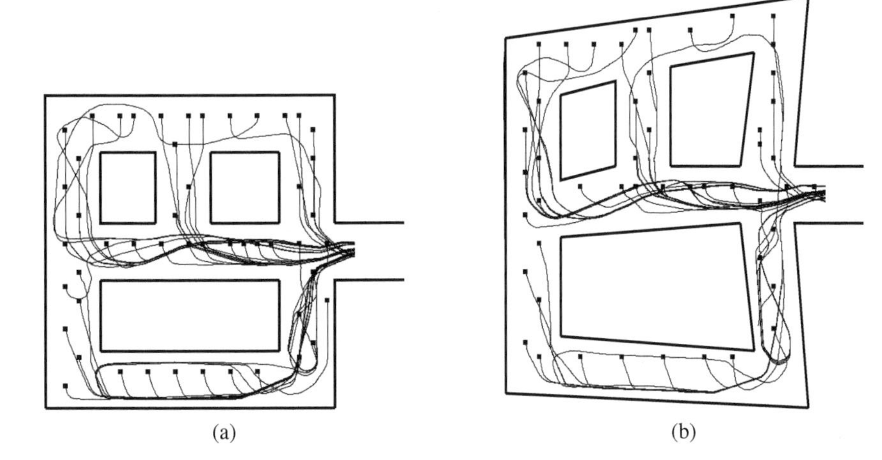

Fig. 7.22: Trajectories of the agent in worlds with modified distances (a) and deformed corridor angles (b): Strategies learned in the unmodified world can be applied successfully. The small squares mark starting positions. Wall colors are not depicted here

Table 7.2: Learned policies also succeed in modified environments: Goal finding success of policies learned in the environment in Fig. 7.17 with TSTC applied in the environments in Fig. 7.22

	Goal finding success	
Representation	Fig. 7.22a	Fig. 7.22b
no TSTC	94 %	88 %
TSTC	100 %	98 %

7.4.2 Policy Transfer to New Environments

In this section we take a look at the transferability of learned policies based on RLPR for application in unknown environments. An agent learned to solve the goal finding task in the world in Fig. 7.3 using le-RLPR with five scans and RLPR_1. In the following, we refer to this as the "source task," and the tasks in new environments where the knowledge is transferred to as "destination tasks."

Both ad hoc transfer (Sect. 5.3.4) for policies learned with task space tile coding (TSTC, Sect. 5.3.3) and a posteriori structure space aspectualization (APSST, Sect. 5.4) were examined. For ad hoc transfer, the learned policies in the source task were applied unmodified to the new scenarios, where also le-RLPR was used for representation. For APSST, structure space policies were derived according to Algorithm 2 such that only structure space was represented by RLPR in the destination

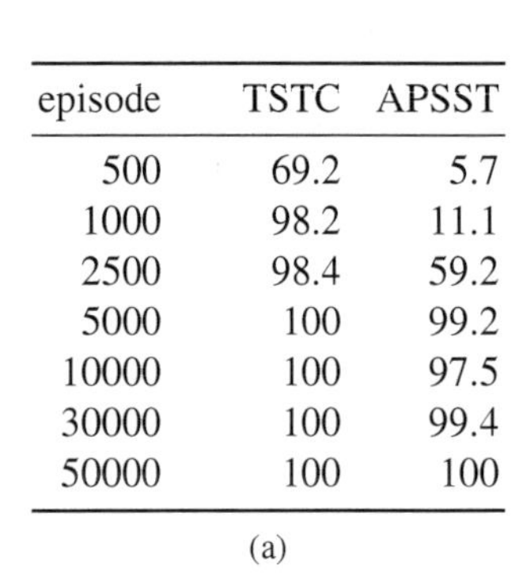

episode	TSTC	APSST
500	69.2	5.7
1000	98.2	11.1
2500	98.4	59.2
5000	100	99.2
10000	100	97.5
30000	100	99.4
50000	100	100

(a)

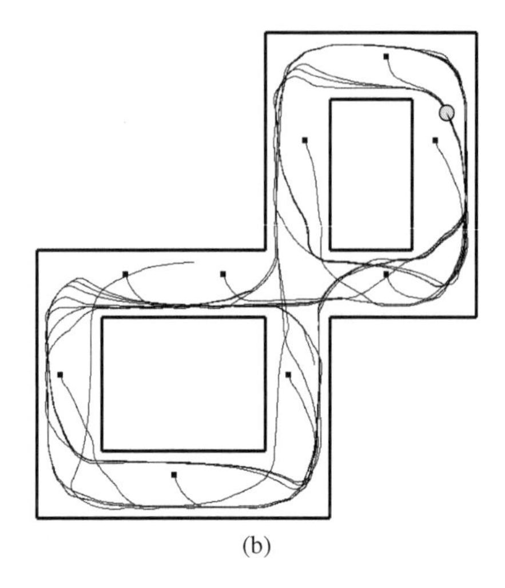

(b)

Fig. 7.23: Transfer performance: Percentage of test runs with successful navigation under transferred policies after various numbers of training episodes with ad hoc and a posteriori transfer (a). The depicted trajectories of the agent in the new environment are the result of ad hoc transfer with TSTC (b)

tasks. Note that in all the experiments described from now on we *only* operate on structure space in the destination tasks. As those operate on RLPR, no landmark observations are included there, and also no goal area is defined within the new worlds. Furthermore, no learning takes place there anymore.

For the experiments, the agent started from 1,000 starting positions spread over the corridors. A test run in the destination task was regarded as successful if the robot was able to navigate within the corridors without colliding for a certain number of time steps. This number was chosen approximately such that the robot could explore the environment twice in such a test run. Test runs in the destination tasks have been performed after a varying number of learning episodes ranging from 500 to 50,000.

Figure 7.23 shows the transfer performance in a circular regular world that is different from the environment of the source task, but does not introduce new structures that have not been perceived during learning. TSTC-based policies performed well in the destination task almost from the start: 69.2% of the test runs after 500 episodes of learning resulted in collision-free navigation in the destination task, and 98.2% after 1,000 episodes. APSST-based policies have comparable success, but take longer: After 500 episodes, only 5.7% of the test runs do not collide, but after 5,000 episodes, APSST also leads to a success rate of 99.2%.

Figures 7.24 and 7.25 show the result in two more complex environments that also include structural elements not present in the source task, for example, dead ends and turns of different angles. Ad hoc transfer with TSTC also had good success rates quite early in these more complicated destination tasks, but did not reach the stable 100% success that was observed in the regular world in Fig. 7.23. While learn-

episode	TSTC	APSST
500	10.6	0.5
1000	26.7	8.4
2500	39.7	62.2
5000	40.6	94.3
10000	62.6	77.5
15000	53.4	85.4
30000	84.4	99.7
50000	98.2	99.9

(a)

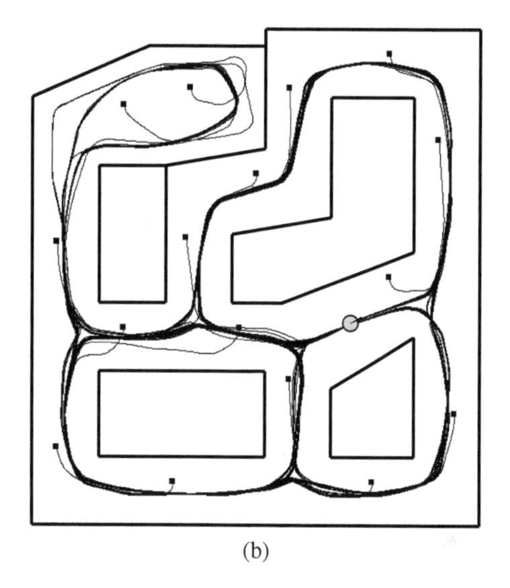

(b)

Fig. 7.24: Percentage of test runs with successful navigation under transferred policies (a). The depicted trajectories of the agent in the new environment are the result of a posteriori structure space transfer (b)

episode	TSTC	APSST
500	5.7	0.0
1000	33.7	0.2
2500	52.2	31.5
5000	37.0	58.7
10000	73.5	97.2
15000	81.8	42.5
30000	84.8	82.0
50000	87.3	96.2

(a)

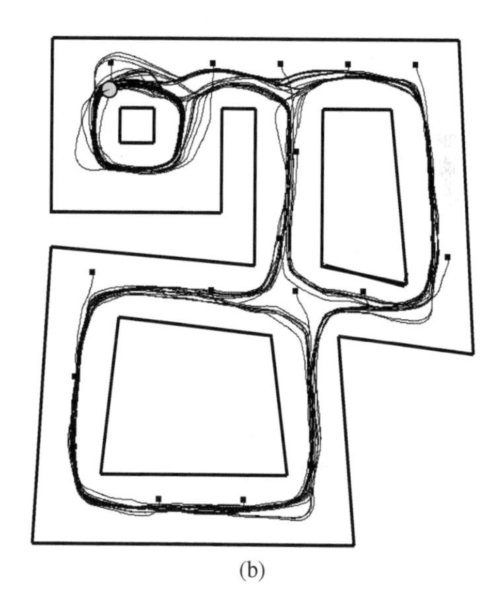

(b)

Fig. 7.25: Percentage of test runs with successful navigation under transferred policies (a). The depicted trajectories of the agent in the new environment are the result of a posteriori structure space transfer (b)

ing proceeds, a posteriori transfer shows a better performance than ad hoc transfer and reaches success rates near 100% even when TSTC-based transfer fails to achieve that.

A posteriori structure space transfer cannot provide successful structure space policies as early as TSTC can. This is because of two reasons: First, learning in the source task did not benefit from TSTC, so the learning was much slower (compared to the results in Sect. 7.2.4). Second, Algorithm 2 needs information in the Q-function for all actions of a state-action pair before it averages over the Q-values and generates new knowledge, while TSTC does not have this restriction.

Summed up, both ad hoc transfer and a posteriori structure space transfer achieve a generally sensible navigation behavior in all the destination tasks. Ad hoc transfer produces a good result early while APSST produces stabler policies that are more successful in the end. Both methods are shown to be suitable for generating policies that can serve as a starting point for learning in unknown environments.

7.5 RLPR-Based Navigation in Real-World Environments

This section describes how le-RLPR can be utilized to perform a very difficult form of knowledge transfer: the transfer of a policy learned in the simulator to a real robotics platform.[5] Because the system dynamics of a simulator and the real world are different and the real world shows much higher complexity, transferring knowledge from a simulator to a real application usually requires a lot of effort (Gabel et al., 2006).

RLPR omits a large part of the environmental details of the world (see Sect. 6.3.4) and offers a highly generalizing view of the world according to the relevant data for action selection. That means that *structurally similar* situations will lead to the same spatial representations even in different environments. This section will show that this also holds for the case where the environment is not a simplified simulation but a physical spatial environment with a real robot operating in it. In particular, this case study shows how a real robot can successfully use a simulator-based structure space policy in a real office environment (see Fig. 7.26).

7.5.1 Properties of a Real Office Environment

The simulator, as introduced in Sect. 7.1.1, only provides straight lines to model the environment the agent is operating in. In real-world office environments, we never encounter such simple structures. Walls are not perfectly straight; they have niches and skirting boards; doors may be open, closed, or half closed; and so on.

[5] This is a revised description of case study that has been presented first in Frommberger and Wolter (2008).

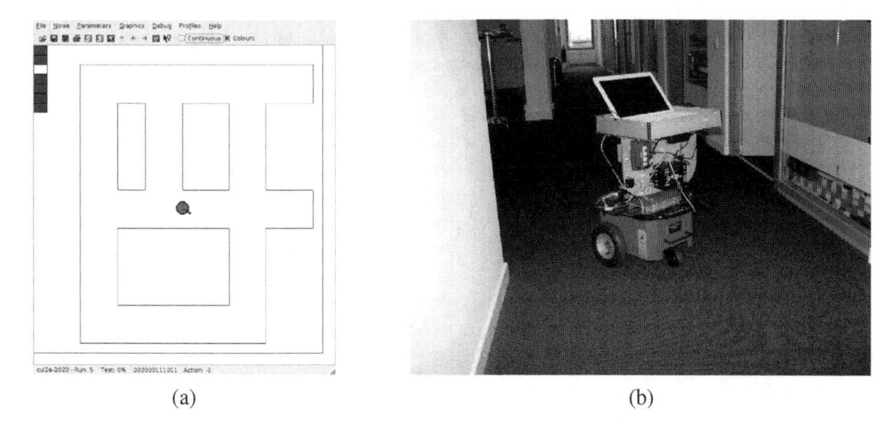

(a) (b)

Fig. 7.26: A screenshot of the robot navigation scenario in the simulator where the strategy is learned (a) and a Pioneer-2 robot in an office building, where the strategy is applied (b). The real office environment offers structural elements not present in the simulator: open space, uneven walls, tables, and other obstacles

Furthermore, there are obstacles, for example, furniture such as cupboards, chairs, and tables. Some of these obstacles are likely to be moved around frequently.

Additionally, we must expect to encounter so-called *dynamic obstacles*, that is, objects that change place while the agent is moving within a scene. Dynamic obstacles can be doors that open or close, humans moving in the corridors, or even other robots operated by fellow colleagues. These objects are usually hard to cope with, because they can hardly be integrated into maps, and strategies to navigate around them are hard to learn due to missing knowledge of the behavior of dynamic obstacles.

In the simulator, we know the exact whereabouts of the line segments, while they have to be detected with the help of sensors on the real robot. Even with the noise model in the simulator (see Sect. 7.3.2.2), one can never exactly model a realistic sensory distortion and uncertainty in measurements. Even the most thorough noise model can never take all the factors into account that are apparent in reality. The real world is pretty different from the simulator—even if it is structurally similar.

7.5.2 Differences of the Real Robot

The robot used for the case study in this work is a Pioneer-2 platform (see Fig. 7.26b) equipped with a laser range finder. It is a wheeled robot with two differential motors driving the front wheels. A laptop computer is mounted on the robot to run the control software.

The properties of the agent in the simulation and the Pioneer-2 platform used for the case study differ significantly in many respects. While the simulated robot is a circle with a certain extent, the Pioneer-2 platform used for the test case has a different shape. The laser range finder is not mounted in the middle of the real robot, so, in contrast to the simulation, information about the walls exactly to the left and the right cannot be achieved. The motion of the real robot is subject to distortion, caused by slippage, uneven ground, or other environmental factors that are not covered by the noise model in the simulator. The most important difference is that the real robot does not execute discrete motion, which is described in the following.

7.5.2.1 Adapting Atomar Actions for a Differential Wheel Drive

In the simulator, the actions the agent could take were small, discrete relocations with regard to coordinates or orientation. These do not comply with the motion behavior of a differential wheel drive, where the motors cannot be operated in a way that allows for such small movement. So, rather than implementing this stepwise motion on the real robot, we map the actions to commands controlling the wheel speeds in order to obtain continuous motion. Additionally, movement is smoothed by averaging the most recent wheel speed commands to avoid strong acceleration/deceleration, which the robot drive cannot handle well. Therefore, averaging was applied to the last eight actions. Given that, we have an action command (v_t, w_t) at a point in time t, where v_t and w_t denote the speed the two motors should perform at according to the requested action primitive. Then the real speed command we send to the motor controller at this time is

$$\frac{1}{8}\sum_{i=0}^{7}(v_{t-i}, w_{t-i}). \tag{7.1}$$

Taking into account that we have a 0.25 second interval for sending wheel commands, this yields a time of two seconds before reaching the wheel speed associated with the action primitive derived from the policy.

7.5.2.2 Adapting the RLPR Grid

As stated in Sect. 6.3.1, the RLPR grid is a property of the agent. Thus, it has to be adapted to the motion dynamics and size of any individual robot. In accordance with that, the immediate surroundings of the RLPR grid have been set to 60 cm in front and 30 cm both to the left and the right of the robot.

Also, we have to choose an appropriate grid layout for the sensor capabilities of the robot. In contrast to the simulator, which assumes (up to) a 360° view, our testbed is only equipped with a 180° laser range finder. The center of the scan is not the center of the robot. Thus, the center of the scan has to be relocated some centimeters back to have it centered with the RLPR grid.

As we do not have any sensory information at the back of the robot, the regions of the RLPR grid that correspond to these areas never contain any data, so they can be omitted. Figure 7.5 shows RLPR_5, the variant of RLPR that takes the mentioned limitations into account and has been used in this experiment.

7.5.3 Operation on Identical Observations

The key issue in knowledge transfer is a shared structure space. To achieve that, both the simulated and the real robot should operate on RLPR. While this comes easily in the simulator, we must make some effort on the real system.

Laser range finders detect obstacles around the robot. By measuring laser beams reflected by obstacles one obtains a sequence of (in our case) 361 points in local coordinates. To abstract from those points to line segments the well-known iterative split-and-merge algorithm that is commonly used in robotics was used to fit lines to scan data (see Gutmann et al. (2001)). As we have seen in Example 4.4, this is a conceptual classification.

As shown in the experiments in Sect. 6.5.2, an overly precise line fitting is not important. More elaborate approaches to extract lines or shapes from a scene, such as Latecki and Lakämper (2000), are not necessary for the given purpose. With respect to parameters, we choose them in a way to make sure that all obstacles detected by the laser range scanners get represented by lines, even if this is a crude approximation. Line prolongation, as suggested in Sect. 6.5.2, has not been used in this experiment.

The detected line configuration is then mapped every 0.25 seconds to the RLPR grid such that we receive an RLPR representation of the surroundings of the real robot (see Fig. 7.27 for a screenshot of the robot control software showing sensory data within an RLPR grid). Now, the robot operates on the same spatial representation as the structure space policy derived from a learned policy in a simulator. Figure 7.28 summarizes the flow of abstraction in the simulator and real environments.

7.5.4 Training and Transfer

The system was trained in the environment depicted in Fig. 7.26a, using le-RLPR with seven color scans and RLPR_5 as described above. The parameters chosen were an input noise level $\rho = 5$ and a motion noise level $\zeta = 0.2$ for training, with a step size $\alpha = 0.05$, a discount factor $\gamma = 0.98$, and $\lambda = 0.9$. The system learned for $40,000$ episodes.

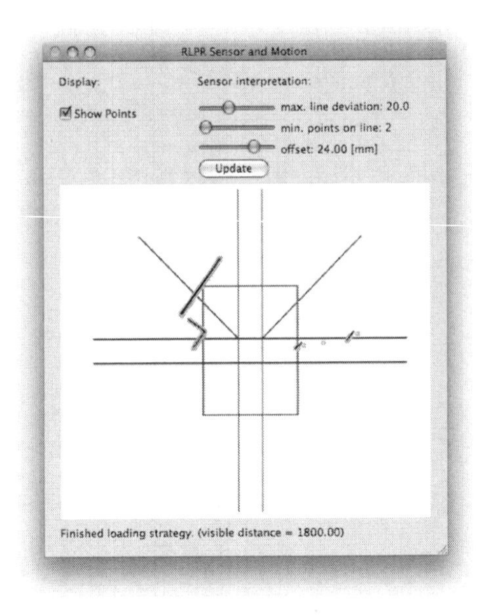

Fig. 7.27 Screenshot: Abstraction to RLPR in the robot controller. Depicted are the RLPR grid, data from the laser range finder (green points), and fitted line segments (red). This output refers to the robot's position shown in Fig. 7.26b

A structure space policy was derived using a posteriori structure space transfer (see Sect. 5.4). This policy could be instantaneously used by the robot control software on the Pioneer-2 platform to enable for navigation in the office environment.

7.5.5 Behavior of the Real Robot

The office building the Pioneer-2 robot had to cope with shows spatial structures that were not present in the simulation. For example, neither open spaces (that is, locations where walls are not present in the immediate view) nor non-rectangular structures (such as big flower pots that we found in the office floor at the time of the experiment) were observed in training. Also, the rooms were equipped with furniture such as sofas and bistro tables.

Figures 7.29 and 7.30 give an impression of the robot experiment: The robot shows reasonable navigation behavior, following corridors in a straight line and turning smoothly around curves. It also shows ability to cope with the structural elements not present in the simulated environment, such as open space or smaller obstacles: All of those are generalized to line segments and thus regarded as if they were walls. This also holds for dynamic obstacles, for example, humans: If they do not move too fast, the robot is able to detect them and to drive around them without collisions. All general navigation skills learned in simulation have been transferred to the real-world environment.

The robot only got stuck when reaching areas with a huge amount of clutter (such as hanging leaves of plants) and in dead ends where the available motion primitives

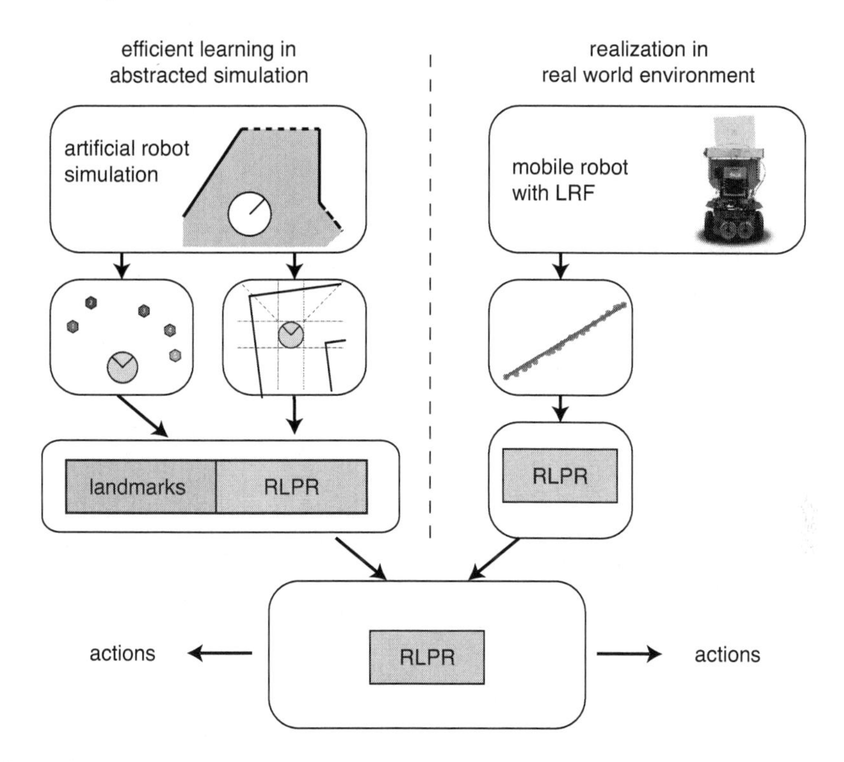

Fig. 7.28: Evolution of spatial representations in both simulation and real robot application. Abstraction techniques enable both scenarios to operate on the RLPR representation to achieve a reasonable action selection

do not allow for collision-free movement anymore. Because the original task was goal-oriented (searching for a specific place), the robot also showed a strong tendency of moving forward and thus actively exploring the environment instead of just avoiding obstacles. This generally sensible navigation behavior could now, for example, be used as a basis for learning new goal-oriented tasks directly on the robotics platform.

7.6 Summary

This section presented an evaluation of the performance of the structure space aspectualizable state space representation le-RLPR in a goal-directed indoor navigation task with Q-learning. Compared to coordinate-based metrical approaches representing the original MDP, the use of le-RLPR allows us to learn successful policies considerably faster with fewer collisions in training. The number of landmark scans had minor influence on the learning performance. The RLPR grid variants $RLPR_0$,

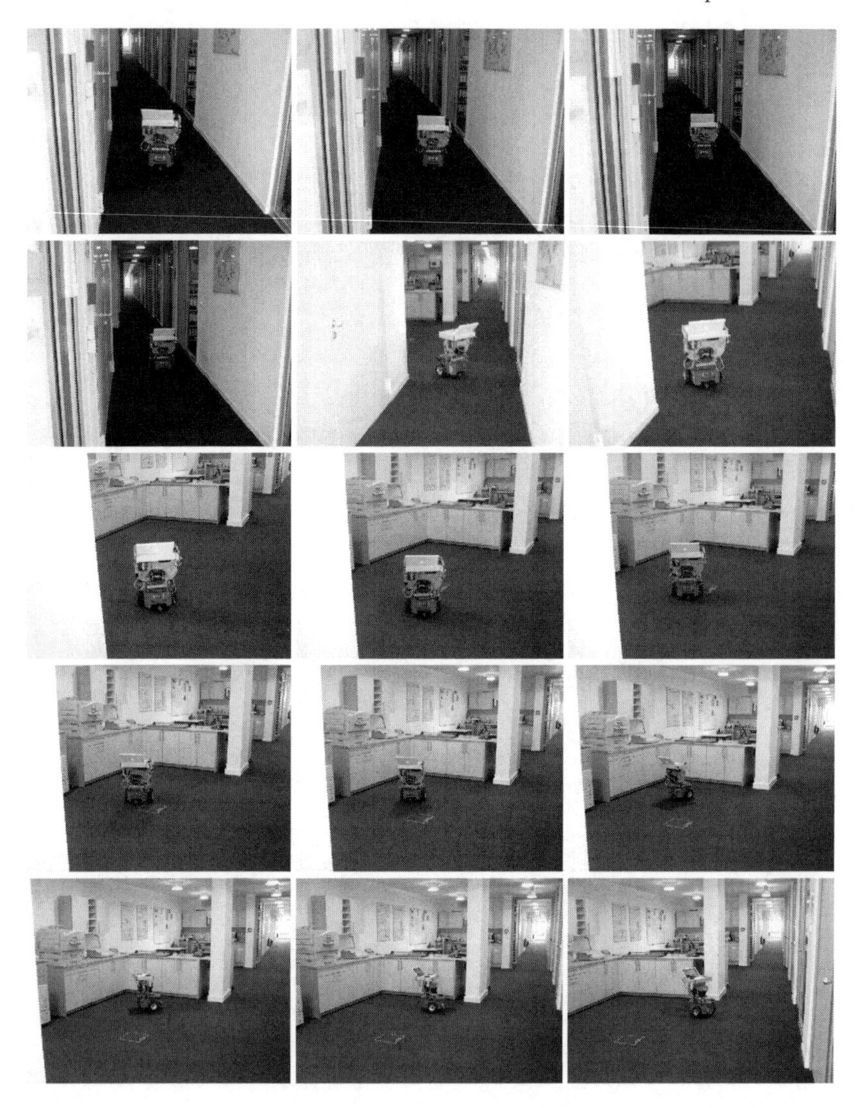

Fig. 7.29: Pioneer-2 follows a corridor and explores an open space, using the RLPR-based structure space policy learned in the simulator. It shows a reasonable navigation behavior in a real office environment, driving smoothly forward and safely around obstacles

$RLPR_1$, and the reduced variant $RLPR_5$ all worked successfully, with advantages for $RLPR_1$ and $RLPR_5$ that split up the regions in the front right and front left. The travel distance of the agent with RLPR only increases between 1.5% and 2.5% with $RLPR_1$ compared to a fine-grained coordinate representation.

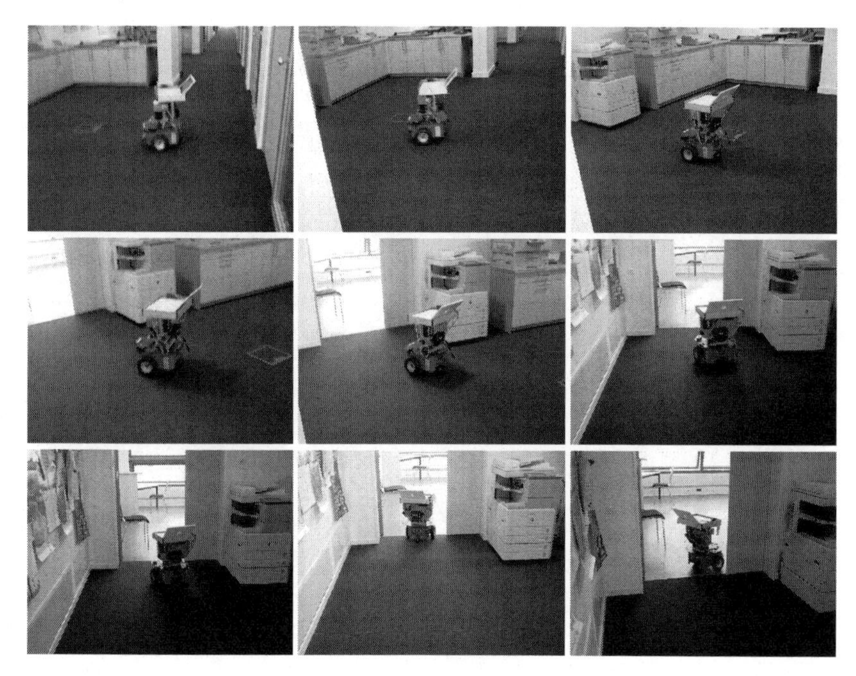

Fig. 7.30: Pioneer-2 leaving an open space through a door to enter the next room

Learning also works without using RLPR and just relying on the sequence of detected landmarks. However, the performance is worse compared to le-RLPR. It has also been shown that le-RLPR can be used with point-based landmarks.

Task space tile coding (TSTC) achieves a powerful non-local generalization and dramatically increases learning speed and helps derive a generally sensible navigation behavior after just a few dozen learning episodes. Also, the use of structure-induced task space aspectualization (SITSA) for landmark selection shows a significant increase in learning speed especially in the early learning phase.

Policies learned with TSTC were successfully applied to modified environments. Furthermore, they were transferred by ad hoc policy transfer to completely new worlds and showed a generally sensible behavior there. A posteriori structure space transfer (APSST) derived structure space policies successfully applicable to unknown environments. Compared to ad hoc transfer with TSTC, those policies needed more training to be successful, but showed better performance in the long run.

le-RLPR proved to be robust against motion noise as well as distorted landmark or line segment detection. This enables us to also transfer policies learned in the simple simulator to a real Pioneer-2 robot platform. Laser range finder data has been interpreted as line segments. This allows us to have an RLPR representation of the real world and to apply an RLPR-based structure space policy immediately.

The robot shows a reasonable navigation behavior in a real office environment, also coping well with unknown structures and even dynamic obstacles.

Chapter 8
Summary and Outlook

This chapter concludes this book with a summary and discussion of the presented work and an outlook on future research based on the described results.

8.1 Summary of the Results

Abstraction is one of the key capabilities of human and artificial agents. It enables us to build concepts from overly rich perceptions in order to cope with complex situations.

Consistent formal frameworks for a computer scientist's view on abstraction were missing in the literature so far. This book provided a formal theory of abstraction and distinguished between three of its facets:

- *Aspectualization* explicates a set of features in a representation by selecting them while implicitly ignoring the others.
- *Coarsening* reduces the granularity of the state space by lowering the number of instances a feature can take.
- *Conceptual Classification* groups semantically related features to form new abstract entities.

In the literature, aspectualization and coarsening are often seen as being conceptually different. In contrast to that it could be shown that based on the formal framework introduced in this book aspectualization and coarsening are equally expressive. Bijections can be defined in source and target space such that an aspectualization can be expressed by a coarsening, and vice versa. This proves that the conceptual distinction between those abstraction facets does not hold for the result of the abstraction. However, the process of aspectualization is to be preferred, as it is computationally simple and features are easily accessible, which allows for coping even with high-dimensional state spaces.

For learning strategies in complex state spaces, the use of spatial abstraction has proved to be an appropriate means. This transforms a Markov decision process

Cognitive Technologies, DOI 10.1007/978-3-642-16590-0_8,

(MDP) into a partially observably Markov decision process (POMDP). However, this POMDP can be solved as if it were an MDP. Of course, the success of this approach depends on how tightly the dynamics of the POMDP and underlying MDP are coupled, that is, on the quality of abstraction with regard to the optimal policy.

To tackle that, abstraction has been identified as an action-driven process. In agent control tasks, the goal of abstraction must be to subsume system states that require the same selection of actions to achieve an optimal policy. To formalize this, the concept of π^*-preservation was described to describe how closely an abstraction can retain optimality of a strategy. A high π^*-preservation quota has been worked out as a central goal of abstraction along with high generalization and high accessibility of the state space representation.

These criteria are most satisfactorily matched by qualitative spatial representations, because they are grounded in solid domain knowledge and emphasize details that make a difference in action selection. In particular, these representations model the structure of a state space. Thus, qualitative spatial abstraction allows for efficient application of reinforcement learning in complex and continuous state spaces.

To benefit from reuse of knowledge in a learning task, this book defined structurally similar state spaces and tasks, and the importance of similar structures for knowledge transfer and generalization was pointed out. For policies, two aspects of behavior can be distinguished:

- *Goal-directed behavior* is task-specific behavior that relates to the goal of the task. It operates on task space.
- *Generally sensible behavior* is task-independent behavior that is valid in all structurally similar tasks. It operates on structure space.

Generally sensible behavior can be reused within and across tasks and environments. Thus, it needs to be represented in an accessible way, and aspectualization proved to be the most suitable facet of abstraction for that. Aspectualizability of a representation has been identified as a core concept for the use of abstraction in agent control tasks. In particular, it enables forming structure space aspectualizable observation spaces that allow for easy access of task and structure space representations.

Two methods for achieving knowledge reuse in structure space aspectualizable state spaces have been introduced in this book:

- *Task space tile coding (TSTC)* applies the CMAC value function approximation to a structure space aspectualizable state space. As an effect, learned knowledge in one state affects all states with an identical structure space descriptor. This enables non-local generalization within a task. Furthermore, ad hoc transfer of a learned policy to a new environment is possible without additional computational effort.
- *A posteriori structure space transfer (APSST)* extracts a structure space policy from a learned goal-directed policy by averaging over Q-values. A structure space policy achieved like this allows for stable transfer to new environments as a basis for new learning tasks.

Furthermore, the confidence of a structure space policy with regard to certainty in action selection can be measured. A similar procedure can also be used to identify non-decision states, which establishes the basis for so-called structure-induced task space aspectualization (SITSA). This method allows for omitting task space information where it is not needed and thus speeds up learning considerably. In a robot navigation context, SITSA has been shown to be suitable for abstracting from landmark information based on the structure of the task.

Tailored for indoor robot navigation tasks, a structure space aspectualizable state space representation called le-RLPR was introduced. It consists of two parts:

- *Ordering information on landmarks* is used to model task space. Two different egocentric variants are regarded: mappings of point-based landmarks to sectors around the robot or samples of views.
- *Relative line position representation (RLPR)* describes the structure of walls relative to the agent's moving direction in an extremely compact way. RLPR shows a high amount of abstraction from details not necessary for generally sensible behavior.

le-RLPR proves to be an excellent abstraction for indoor robot navigation tasks. It shrinks the state space considerably and outperforms coordinate-based approaches in empirical tests. When combined with task space tile coding, le-RLPR shows a rapid learning due to non-local generalization.

Empirical evaluation also shows that le-RLPR is suitable for both ad hoc and a posteriori knowledge transfer. Structure space policies derived from learned policies in the source task are applicable to new tasks in completely new environments.

Also, le-RLPR is very robust with regard to environmental noise. This allows using this representation in combination with APSST for transferring a policy learned in a simple simulator to a real robotics platform. The emerging structure space policy allows a Pioneer-2 robot to safely navigate and explore a real office environment.

The work described in this book contributes to several research areas:

- For the machine learning community it shows how appropriate use of abstraction paradigms and the choice of a state space representation tailored for the task at hand influences important performance indicators beneficially and scales down hard problems significantly. Furthermore, it introduces the concept of structure space aspectualizable representations, which are the key to knowledge reuse within and across tasks.
- For the spatial cognition community it contributes a formal framework of abstraction for knowledge representation and important insights into the nature of different facets of abstraction. It also offers successful application of qualitative spatial representations to a real-world problem that leads to clearly visible performance improvement.
- To the field of robotics this work presents a very compact world model that contrasts with the usually employed huge stochastic approaches; but it allows for successful robot navigation and even can bridge the gap from simulation to real robotics.

The presented theoretical considerations and empirical findings prove that aspectualizable qualitative spatial abstraction is a fundamental concept to for enabling reinforcement learning to efficiently solve tasks in complex and continuous state spaces and allows for effective knowledge reuse in different problems.

8.2 Future Work

The work presented in this book can be a basis for further research in the field of qualitative spatial abstraction in reinforcement learning:

- Structure space policies have proved to be suitable for achieving a generally sensible behavior of an agent in new tasks and, thus, can be a basis for further learning activities. Task space tile coding and a posteriori structure space transfer have proved to be appropriate methods for creating structure space policies. The question of how exactly their results can be integrated into the new learning process and how this can be utilized to achieve life-long learning still has to be investigated in detail. Also, the hierarchical approach to learn task space and structure space policies separately should be investigated and evaluated in greater detail.
- The RLPR grid is a property of the agent depending on its motion dynamics. At the moment, it has to be given a priori. A rewarding possibility would be to try to adapt the RLPR grid prior to or while learning, such that structure space policies can be transferred to different platforms (for example, different robot models) even more easily.
- le-RLPR in conjunction with task space tile coding has shown rapid learning progress that could even allow for learning on real hardware instead of the simulator. This is an interesting field for further investigation.
- The presented work concentrates on the transfer of structure space knowledge. However, parts of task space information can be worth transferring to new tasks as well in certain situations. Therefore, it is necessary to identify these situations and to provide mechanisms for task space transfer.
- Qualitative spatial representations provide a solid symbol grounding that especially allows for interaction with humans. For example, route knowledge given in natural language can be integrated as background knowledge into the learning process. RLPR can be used to identify structures within the environment and relate them to natural language. A study on identification of spatial structure based on RLPR has already been carried out in a master's thesis (Khan, 2009) that links RLPR-based navigation strategies to topological mapping.
- Identification of decision and non-decision points is used to provide situation-dependent abstraction in the scope of this book. It could also be used for additional purposes, for example, to ensure a better exploration at places that appear to be important according to their structure.
- The identification of structure space is a critical issue for building structure space aspectualizable representations. This work assumes that knowledge about structure space arises from background knowledge and becomes part of the design of

the learning task. It would be most rewarding to have methods to detect structure space or parts of it automatically. This would be an important step towards agents that could autonomously reuse previously gained knowledge when confronted with a completely new situation. In the author's view, investigating autonomous structure space detection is the most important challenge brought up by the work presented in this book.

References

Al-Ansari, M.A., Williams, R.J.: Robust, efficient, globally-optimized reinforcement learning with the parti-game algorithm. In: Kearns, M.S., Solla, S.A., Cohn, D.A. (eds.) Advances in Neural Information Processing Systems: Proceedings of the 1998 Conference, pp. 961–967. MIT Press, Cambridge, MA (1999)

Albus, J.S.: Brain, Behavior, and Robotics. Byte Books, Peterborough, NH, USA (1981)

Amarel, S.: On representations of problems of reasoning about actions. Machine Intelligence **3**, 131–171 (1698). Reprint in Webber and Nilsson (1981)

Andre, D., Russell, S.J.: State abstraction for programmable reinforcement learning agents. In: Proceedings of the National Conference on Artificial Intelligence (AAAI), pp. 119–125 (2002)

Arleo, A., del R. Millán, J., Floreano, D.: Efficient learning of variable-resolution cognitive maps for autonomous indoor navigation. IEEE Transactions on Robotics and Automation **15**(6) (1999)

Aström, K.J.: Optimal control of Markov processes with incomplete state information. Journal of Mathematical Analysis and Applications **10**, 174–205 (1965)

Baird, L.: Residual algorithms: Reinforcement learning with function approximation. In: Proceedings of the Twelfth International Conference on Machine Learning (ICML), pp. 30–37. Morgan Kaufmann, San Francisco, CA (1995)

Barkowsky, T., Berendt, B., Egner, S., Freksa, C., Krink, T., Röhrig, R., Wulf, A.: The realator – how to construct reality. In: Proceedings of the ECAI Workshop on Spatial and Temporal Reasoning, pp. 19–26 (1994)

Barkowsky, T., Berendt, B., Egner, S., Freksa, C., Latecki, L., Röhrig, R.: Movies: an integrated space-time representation for qualitative navigation. In: Krieg-Brückner, B., Herwig, C. (eds.) Tagungsband des Workshops Kognitive Robotik (1995)

Barto, A.G., Mahadevan, S.: Recent advances in hierarchical reinforcement learning. Discrete Event Dynamic Systems **13**(4) (2004)

Barto, A.G., Sutton, R.S., Anderson, C.W.: Neuronlike adaptive elements that can solve difficult learning control problems. IEEE Transactions on Systems, Man, and Cybernetics **8** (1983)

Baum, L.E., Sell, G.R.: Growth functions for transformations on manifolds. Pacific Journal of Mathematics **27**(2), 211–227 (1968)

Bellman, R.E.: Dynamic Programming. Princeton University Press, Princeton, NJ (1957)

Berendt, B., Barkowsky, T., Freksa, C., Kelter, S.: Spatial representation with aspect maps. In: Freksa, C., Habel, C., Wender, K.F. (eds.) Spatial Cognition: an Interdisciplinary Approach to Representing and Processing Spatial Knowledge, *Lecture Notes in Artificial Intelligence*, vol. 1404, pp. 313–336. Springer, Berlin (1998)

Berry, D.A., Fristedt, B.: Bandit Problems: Sequential Allocation of Experiments. Monographs on Statistics and Applied Probability. Chapman and Hall, London, UK (1985)

L. Frommberger, *Qualitative Spatial Abstraction in Reinforcement Learning*,
Cognitive Technologies, DOI 10.1007/978-3-642-16590-0,

Bertel, S., Freksa, C., Vrachliotis, G.: Aspectualize and conquer in architectural design. In: Gero, J.S., Tversky, B., Knight, T. (eds.) Visual and Spatial Reasoning in Design III, pp. 255–279. Key Centre of Design, Computing and Cognition, University of Sydney (2004)

Bertel, S., Vrachliotis, G., Freksa, C.: Aspect-oriented building design: Toward computer-aided approaches to solving spatial contraint problems in architecture. In: Allen, G.L. (ed.) Applied Spatial Cognition: From Research to Cognitive Technology, pp. 75–102. Lawrence Erlbaum Associates, Mahwah, NJ, USA (2007)

Bertsekas, D.P.: Dynamic Programming. Prentice-Hall (1987)

Bianchi, R.A.C., Ribeiro, C.H.C., Costa, A.H.R.: Heuristic selection of actions in multiagent reinforcement learning. In: Proceedings of the Twentieth International Joint Conference on Artificial Intelligence (IJCAI), pp. 690–695. Hyderabad, India (2007)

Bittner, T., Smith, B.: A taxonomy of granular partitions. In: Montello, D. (ed.) Spatial Information Theory: Cognitive and Computational Foundations of Geographic Information Science (COSIT), *Lecture Notes in Computer Science*, vol. 2205, pp. 28–43. Springer, Berlin (2001)

Boyan, J.A., Moore, A.W.: Generalization in reinforcement learning: Safely approximating the value function. In: Tesauro, G., Touretzky, D., Leen, T. (eds.) Advances in Neural Information Processing Systems: Proceedings of the 1994 Conference, vol. 7, pp. 369–376. MIT Press, Cambridge, MA (1995)

Brafman, R.I., Tennenholtz, M.: R-MAX – a general polynomial time algorithm for near-optimal reinforcement learning. Journal of Machine Learning Research **3**, 213–231 (2003)

Braga, A.P.S., Araújo, A.F.R.: A topological reinforcement learning agent for navigation. Neural Computing and Applications **12**, 220–236 (2003)

Busquets, D., de Mántaras, R.L., Sierra, C., Dietterich, T.G.: Reinforcement learning for landmark-based robot navigation. In: Proceedings of the Autonomous Agents and Multiagent Systems Conference (AAMAS), pp. 841–843. Bologna, Italy (2002)

Cassandra, A.R., Kaelbling, L.P., Littman, M.L.: Acting optimally in partially observable stochastic domains. In: Proceedings of the Twelfth National Conference on Artificial Intelligence (AAAI). Seattle, WA (1994)

Cohn, A.G., Hazarika, S.M.: Qualitative spatial representation and reasoning: An overview. Fundamenta Informaticae **46**(1–2), 1–29 (2001)

Crook, P., Hayes, G.: Learning in a state of confusion: Perceptual aliasing in grid world navigation. In: Proceedings of Towards Intelligent Mobile Robots (TIMR). UWE, Bristol (2003)

Dayan, P.: The convergence of TD(λ) for general λ. Machine Learning **8**, 341–362 (1992)

Dean, T., Givan, R.: Model minimization in Markov decision processes. In: Proceedings of the 14th National Conference on Artificial Intelligence (AAAI), pp. 106–111. Providence, RI, USA (1997)

Defense Advanced Research Projects Agency: Transfer learning proposer information pamphlet (PIP) for broad agency announcement 05–29. Available at `http://www.darpa.mil/IPTO/solicit/baa/BAA-05-29_PIP.pdf` (2005)

Degris, T., Sigaud, O., Wuillemin, P.H.: Learning the structure of factored Markov decision processes in reinforcement learning problems. In: Proceedings of the Twenty Third International Conference on Machine Learning (ICML), pp. 257–264. Pittsburgh, PA (2006)

Dietterich, T.G.: The MAXQ method for hierarchical reinforcement learning. In: Proceedings of the Fifteenth International Conference on Machine Learning (ICML), pp. 118–126. Morgan Kaufmann (1998)

Dietterich, T.G.: Hierarchical reinforcement learning with the MAXQ value function decomposition. International Journal of Artificial Intelligence Research **13**, 227–303 (2000a)

Dietterich, T.G.: State abstraction in MAXQ hierarchical reinforcement learning,. In: Solla, S.A., Leen, T.K., Müller, K.R. (eds.) Advances in Neural Information Processing Systems 12: Procedings of the 1999 Conference, pp. 994–1000. MIT Press (2000b)

Dylla, F., Frommberger, L., Wallgrün, J.O., Wolter, D., Nebel, B., Wölfl, S.: SailAway: Formalizing navigation rules. In: Proceedings of AISB Symposium on Spatial Reasoning and Communication. Edinburgh, UK (2007)

Encyclopædia Britannica: Learning. Encyclopædia Britannica Online (http://www.britannica.com/ebc/article-9369902) (2007)
Escrig, M.T., Toledo, F.: Autonomous robot navigation using human spatial concepts. International Journal of Intelligent Systems **15**(3), 165–196 (2000)
Fernández, F., Veloso, M.: Probabilistic policy reuse in a reinforcement learning agent. In: Proceedings of the Fifth International Joint Conference on Autonomous Agents and Multiagent Systems (AAMAS), pp. 720–727. Hakodate, Japan (2006)
Forbus, K.D.: Qualitative Process Theory. Artificial Intelligence **24**, 85–168 (1984)
Frank, A.: Qualitative spatial reasoning about cardinal directions. In: Proceedings of the American Congress on Surveying and Mapping (ACSM-ASPRS), pp. 148–167. Baltimore, Maryland, USA (1991)
Frank, A.: Qualitative spatial reasoning: Cardinal directions as an example. International Journal of Geographical Information Systems **10**(3), 269–290 (1996)
Franz, M.O., Schölkopf, B., Mallot, H.A., Bülthoff, H.H.: Learning view graphs for robot navigation. Autonomous Robots **5**, 111–125 (1998)
Freksa, C.: Using orientation information for qualitative spatial reasoning. In: Frank, A.U., Campari, I., Formentini, U. (eds.) Theories and methods of spatio-temporal reasoning in geographic space, pp. 162–178. Springer, Berlin (1992)
Freksa, C., Röhrig, R.: Dimensions of qualitative spatial reasoning. In: Carreté, N.P., Singh, M.G. (eds.) Proceedings of the 3rd IMACS International Workshop on Qualitative Reasoning and Decision Technologies (QUARDET), pp. 483–492. Barcelona, Spain (1993)
Freksa, C., Zimmermann, K.: On the utilization of spatial structures for cognitively plausible and efficient reasoning. In: Proceedings of the IJCAI Workshop on Spatial and Temporal Reasoning, pp. 61–66. Chambéry, France (1993)
Frommberger, L.: A qualitative representation of structural spatial knowledge for robot navigation with reinforcement learning. In: Proceedings of the ICML Workshop on Structural Knowledge Transfer for Machine Learning. Pittsburgh, PA, USA (2006)
Frommberger, L.: Generalization and transfer learning in noise-affected robot navigation tasks. In: Neves, J.M., Santos, M.F., Machado, J.M. (eds.) Progress in Artificial Intelligence: Proceedings of EPIA 2007, *Lecture Notes in Artificial Intelligence*, vol. 4874, pp. 508–519. Springer-Verlag Berlin Heidelberg, Guimarães, Portugal (2007a)
Frommberger, L.: A generalizing spatial representation for robot navigation with reinforcement learning. In: Proceedings of the Twentieth International Florida Artificial Intelligence Research Society Conference (FLAIRS), pp. 586–591. AAAI Press, Key West, FL, USA (2007b)
Frommberger, L.: Learning to behave in space: A qualitative spatial representation for robot navigation with reinforcement learning. International Journal on Artificial Intelligence Tools **17**(3), 465–482 (2008a)
Frommberger, L.: Representing and selecting landmarks in autonomous learning of robot navigation. In: Xiong, C., Liu, H., Huang, Y., Xiong, Y. (eds.) Intelligent Robotics and Applications: First International Conference (ICIRA 2008), Part I, *Lecture Notes in Artificial Intelligence*, vol. 5314, pp. 488–497. Springer Verlag Berlin Heidelberg (2008b)
Frommberger, L.: Situation dependent spatial abstraction in reinforcement learning based on structural knowledge. In: Proceedings of the ICML/UAI/COLT Workshop on Abstraction in Reinforcement Learning. Montreal, Canada (2009)
Frommberger, L., Wolter, D.: Spatial abstraction: Aspectualization, coarsening, and conceptual classification. In: Freksa, C., Newcombe, N.S., Gärdenfors, P., Wölfl, S. (eds.) Spatial Cognition VI: Reasoning, Action, Interaction: International Conference Spatial Cognition, *Lecture Notes in Artificial Intelligence*, vol. 5248, pp. 311–327. Springer Verlag Berlin Heidelberg (2008)
Gabel, T., Hafner, R., Lange, S., Lauer, M., Riedmiller, M.: Bridging the gap: Learning in the RoboCup simulation and midsize league. In: Proceedings of the 7th Portuguese Conference on Automatic Control (Controlo 2006). Lisbon, Portugal (2006)

Gabel, T., Riedmiller, M.: CBR for state value function approximation in reinforcement learning. In: Proceedings of the Sixteenth European Conference on Machine Learning (ECML), pp. 206–221. Bonn, Germany (2005)
Galton, A., Meathrel, R.C.: Qualitative outline theory. In: Proceedings of the Sixteenth International Conference on Artificial Intelligence (IJCAI), pp. 1061–1066. Stockholm, Sweden (1999)
Gantner, Z., Westphal, M., Wölfl, S.: GQR – a fast reasoner for binary qualitative constraint calculi. In: Proceedings of the AAAI Workshop on Spatial and Temporal Reasoning. Chicago, IL (2008)
Glaubius, R., Namihira, M., Smart, W.D.: Speeding up reinforcement learning using manifold representations: Preliminary results. In: Proceedings of the IJCAI Workshop "Reasoning with Uncertainty in Robotics". Edinburgh, Scotland (2005)
Glaubius, R., Smart, W.D.: Manifold representations for value-function approximation. In: Proceedings of the AAAI Workshop on Markov Decision Processes. San Jose, CA (2004)
Gordon, G.J.: Stable function approximation on dynamic programming. In: Proceedings of the Twelfth International Conference on Machine Learning (ICML), pp. 261–268. Morgan Kaufmann, San Francisco (1995)
Goyal, R.: Similarity assessment for cardinal directions between extended spatial objects. Ph.D. thesis, University of Maine (2000)
Goyal, R.K., Egenhofer, M.J.: Consistent queries over cardinal directions across different levels of detail. In: Tjoa, A.M., Wagner, R., Al-Zobaidie, A. (eds.) Proceedings of the 11th International Workshop on Database and Expert System Applications, pp. 867–880. IEEE Computer Society, Greenwich, UK (2000)
Gutmann, J.S., Weigel, T., Nebel, B.: A fast, accurate and robust method for self-localization in polygonal environments using laser range finders. Advanced Robotics **14**(8), 651–667 (2001)
Hasinoff, S.W.: Reinforcement learning for problems with hidden state. Tech. rep., University of Toronto, Toronto, Canada (2003)
Herskovits, A.: Schematization. In: Olivier, P., Gapp, K.P. (eds.) Representation and Processing of Spatial Expressions, pp. 149–162. Lawrence Erlbaum Associates, Mahwah, NJ, USA (1998)
Hobbs, J.R.: Granularity. In: Proceedings of the Ninth International Joint Conference on Artificial Intelligence (IJCAI), pp. 432–435. Los Angeles, CA, USA (1985)
Howard, R.A.: Dynamic Probabilistic Systems, Volume II: Semi-Markov and Decision Processes. Dover Publications (1971)
Jong, N.K., Stone, P.: State abstraction discovery from irrelevant state variables. In: Proceedings of the Nineteenth International Conference on Artificial Intelligence (IJCAI), pp. 752–757 (2005)
Kaelbling, L.P., Littman, M.L., Cassandra, A.R.: Planning and acting in partially observable stochastic domains. Artificial Intelligence **101**, 99–134 (1998)
Kaelbling, L.P., Littmann, M.L., Moore, A.W.: Reinforcement learning: A survey. Journal of Artificial Intelligence Research **4**, 237–285 (1996)
Khan, M.S.: A hierarchical map-building strategy for mobile robot navigation. Master's thesis, Department of Computer Science and Engineering, Indian Institute of Technology, Kanpur, India (2009). Joint supervision with the Faculty of Mathematics and Informatics, Universität Bremen
Kirchner, F.: Q-learning of complex behaviors on a six-legged walking machine. Journal of robotics and autonomous systems **25**, 256–263 (1998)
Klatzky, R.L.: Allocentric and egocentric spatial representations: Definition, distinctions, and interconnections. In: Freksa, C., Habel, C., Wender, K.F. (eds.) Spatial Cognition. An Interdisciplinary Approach to Representing and Processing Spatial Knowledge, *Lecture Notes in Artificial Intelligence*, vol. 1404. Springer (1998)
Klippel, A., Richter, K.F., Barkowsky, T., Freksa, C.: The cognitive reality of schematic maps. In: Meng, L., Zipf, A., Reichenbacher, T. (eds.) Map-based Mobile Services – Theories, Methods and Implementations, pp. 57–74. Springer, Berlin (2005)

Konidaris, G.D.: A framework for transfer in reinforcement learning. In: Proceedings of the ICML Workshop on Structural Knowledge Transfer for Machine Learning. Pittsburgh, PA, USA (2006)

Konidaris, G.D., Barto, A.G.: Autonomous shaping: Knowledge transfer in reinforcement learning. In: Proceedings of the Twenty Third International Conference on Machine Learning (ICML), pp. 489–49. Pittsburgh, PA (2006)

Konidaris, G.D., Barto, A.G.: Building portable options: Skill transfer in reinforcement learning. In: Proceedings of the Twentieth International Joint Conference on Artificial Intelligence (IJ-CAI) (2007)

Kuipers, B.: The spatial semantic hierarchy. Artificial Intelligence **119**, 191–233 (2000)

Lane, T., Wilson, A.: Toward a topological theory of relational reinforcement learning for navigation tasks. In: Proceedings of the Eighteenth International Florida Artificial Intelligence Research Society Conference (FLAIRS) (2005)

Latecki, L.J., Lakämper, R.: Shape similarity measure based on correspondence of visual parts. IEEE Trans. Pattern Analysis and Machine Intelligence **22**(10), 1185–1190 (2000)

LaValle, S.M.: Planning Algorithms. Cambridge University Press (2006). Available online at `http://planning.cs.uiuc.edu/web.html`

Lazanas, A., Latombe, J.C.: Landmark-based robot navigation. Algorithmica **13**(5), 472–501 (1995)

Levinson, S.C.: Space in Language and Cognition: Explorations in Cognitive Diversity, chap. 2. Cambridge University Press (2003)

Levitt, T.S., Lawton, D.T.: Qualitative navigation for mobile robots. Artificial Intelligence **44**, 305–360 (1990)

Li, L., Walsh, T.J., Littman, M.L.: Towards a unified theory of state abstraction for MDPs. In: Proceedings of the Ninth International Symposium on Artificial Intelligence and Mathematics, pp. 531–539 (2006)

Ligozat, G., Renz, J.: What is a qualitative calculus? A general framework. In: Zhang, C., Guesgen, H.W., Yeap, W.K. (eds.) PRICAI 2004: Trends in Artificial Intelligence, 8th Pacific RimInternational Conference on Artificial Intelligence, Auckland, New Zealand, Proceedings, *Lecture Notes in Computer Science*, vol. 3157, pp. 53–64. Springer (2004)

Likhachev, M., Koenig, S.: Speeding up the parti-game algorithm. In: Becker, S., Thrun, S., Obermayer, K. (eds.) Advances in Neural Information Processing Systems: Proceedings of the 2002 Conference, pp. 1563–1570. MIT Press, Cambridge, MA (2003)

Littman, M.L.: Memoryless policies: Theoretical limitations and practical results. In: Cliff, D., Husbands, P., Meyer, J.A., Wilson, S.W. (eds.) From Animals to Animats 3: Proceedings of the Third International Conference on Simulation of Adaptive Behavior (SAB), pp. 238–245. Brighton, UK (1994)

Littman, M.L., Cassandra, A.R., Kaelbling, L.P.: Learning policies for partially observable environments: Scaling up. In: Proceedings of the Twelfth International Conference on Machine Learning (ICML), pp. 362–370. Morgan Kaufmann, San Francisco (1994)

Littmann, M.L., Dean, T.L., Kaelbling, L.P.: On the complexity of solving Markov decision problems. In: Proceedings of the Eleventh Annual Conference on Uncertainty in Artificial Intelligence (UAI-95). Montreal, Canada (1995)

Liu, Y., Stone, P.: Value-function-based transfer for reinforcement learning using structure mapping. In: Proceedings of the National Conference on Artificial Intelligence (AAAI). Boston, MA (2006)

Lovejoy, W.S.: A survey of algorithmic methods for partially observable Markov decision processes. Annals of Operations Research **1**, 47–66 (1991)

Mackaness, W.A., Chaudhry, O.: Generalization and symbolization. In: Shekhar, S., Xiong, H. (eds.) Encyclopedia of GIS. Springer (2008)

Mahadevan, S.: Samuel meets Amarel: Automatic value function approximation using global state space analysis. In: Proceedings of the National Conference on Artificial Intelligence (AAAI), pp. 877–917. Pittsburgh, PA (2005)

Mahadevan, S., Maggioni, M.: Value function approximation with diffusion wavelets and Laplacian eigenfunctions. In: Weiss, Y., Schölkopf, B., Platt, J. (eds.) Advances in Neural Information Processing Systems: Proceedings of the 2005 Conference, vol. 18, pp. 843–850. MIT Press, Cambridge, MA (2006)

Mallot, H.A., Franz, M., Schölkopf, B., Bülthoff, H.H.: The view-graph approach to visual navigation and spatial memory. In: Proceedings of the 7th International Conference on Artificial Neural Networks (ICANN), ZKW Bericht. Zentrum für Kognitionswissenschaften, Universität Bremen, Lausanne (1997)

McCallum, A.K.: Reinforcement learning with selective perception and hidden state. Ph.D. thesis, Department of Computer Science, University of Rochester, NY (1995)

McGovern, A., Barto, A.G.: Autonomous discovery of temporal abstract actions from interaction with an environment. In: Proceedings of the Eighteenth International Conference on Machine Learning (ICML), pp. 361–368. Morgan Kaufmann, San Francisco, CA (2001)

Moore, A.W., Atkeson, C.G.: Prioritized sweeping: Reinforcement learning with less data and less time. Machine Learning **13**, 103–130 (1993)

Moore, A.W., Atkeson, C.G.: The parti-game algorithm for variable resolution reinforcement learning in multidimensional state-spaces. Machine Learning **21**(3), 199–233 (1995)

Moratz, R.: A granular point position calculus. Tech. Rep. 005, Cognitive Systems – SFB/TR8 Spatial Cognition, Universität Bremen (2005)

Moratz, R.: Representing relative direction as binary relation of oriented points. In: Proceedings of the 17th European Conference on Artificial Intelligence (ECAI). Riva del Garda, Italy (2006)

Moratz, R., Dylla, F., Frommberger, L.: A relative orientation algebra with adjustable granularity. In: Proceedings of the Workshop on Agents in Real-Time and Dynamic Environments (IJCAI 05). Edinburgh, Scotland (2005)

Moravec, H.P., Elfes, A.E.: High resolution maps from wide angle sonar. In: Proceedings of the IEEE International Conference on Robotics and Automation (ICRA). St. Louis, MO (1985)

Mukerjee, A., Joe, G.: A qualitative model for space. In: Proceedings of the Eighth National Conference on Artificial Intelligence (AAAI), pp. 721–727. Morgan Kaufmann, Boston, MA (1990)

Munos, R., Moore, A.: Variable resolution discretizations for high-accuracy solutions of optimal control problems. In: Proceedings of the Sixteenth International Conference on Artificial Intelligence (IJCAI), pp. 1348–1355. Stockholm, Sweden (1999)

Musto, A., Stein, K., Schill, K., Eisenkolb, A., Brauer, W.: Qualitative motion representation in egocentric and allocentric frames of reference. In: Freksa, C., Mark, D.M. (eds.) Spatial Information Theory. Cognitive and Computational Foundations of Geographic Information Science (COSIT), vol. 1661, pp. 461–476. Springer, Berlin (1999)

Nilsson, N.J.: Introduction to machine learning: Draft of incomplete notes. Published online at `http://ai.stanford.edu/people/nilsson/mlbook.html` (1996)

Owen, C., Nehmzow, U.: Landmark-based robot navigation. In: From Animals to Animats 5: Proceedings of the Fifth International Conference on Simulation of Adaptive Behavior (SAB). Zurich, Switzerland (1998)

Parr, R., Russell, S.J.: Reinforcement learning with hierarchies of machines. In: Jordan, M.I., Kearns, M.J., Solla, S.A. (eds.) Advances in Neural Information Processing Systems 10: Procedings of the 1997 Conference, pp. 1043–1049. MIT Press (1998)

Peng, J., Williams, R.J.: Incremental multi-step Q-learning. In: Proceedings of the Eleventh International Conference on Machine Learning (ICML), pp. 226–232. Morgan Kaufmann, San Francisco (1994)

Peng, J., Williams, R.J.: Incremental multi-step Q-learning. Machine Learning **22**, 283–290 (1996)

Petrik, M.: An analysis of Laplacian methods for value function approximation in MDPs. In: Proceedings of the Twentieth International Joint Conference on Artificial Intelligence (IJCAI), pp. 2574–2579. Hyderabad, India (2007)

Porta, J.M., Celaya, E.: Reinforcement learning for agents with many sensors and actuators acting in categorizable environments. Journal of Artificial Intelligence Research **23**, 79–122 (2005)

Powell, M.J.D.: Radial basis functions for multivariate interpolation: A review. In: Mason, J.C., Cox, M.G. (eds.) Algorithms for Approximation, The Institute of Mathematics and its Applications Conference Series, pp. 143–167. Clarendon Press, Oxford (1987)

Precup, D.: Temporal abstraction in reinforcement learning. Ph.D. thesis, Department of Computer Science, University of Massachusetts, Amherst MA (2000)

Prescott, T.J.: Spatial representation for navigation in animats. Adaptive Behavior **4**(2), 85–125 (1996)

Presson, C.C., Montello, D.R.: Points of reference in spatial cognition: Stalking the elusive landmark. British Journal of Developmental Psychology **6**, 378–381 (1988)

Ravindran, B.: An algebraic approach to abstraction in reinforcement learning. Ph.D. thesis, Department of Computer Science, University of Massachusetts, Amherst MA (2004)

Ravindran, B., Barto, A.G.: SMDP homomorphisms: An algebraic approach to abstraction in semi-Markov decision processes. In: Proceedings of the Sixteenth International Joint Conference on Artificial Intelligence (IJCAI), pp. 1011–1018. Acapulco, Mexico (2003)

Renz, J., Mitra, D.: Qualitative direction calculi with arbitrary granularity. In: Zhang, C., Guesgen, H.W., Yeap, W.K. (eds.) PRICAI 2004: Trends in Artificial Intelligence, 8th Pacific RimInternational Conference on Artificial Intelligence, Auckland, New Zealand, Proceedings, *Lecture Notes in Computer Science*, vol. 3157, pp. 65–74. Springer (2004)

Reynolds, S.I.: Adaptive resolution model-free reinforcement learning: Decision boundary partitioning. In: Proceedings of the Seventeenth International Conference on Machine Learning (ICML). Morgan Kaufmann, San Francisco (2000)

Richter, K.F.: A uniform handling of different landmark types in route directions. In: Winter, S., Duckham, M., Kulik, L., Kuipers, B. (eds.) Spatial Information Theory, LNCS 4736, pp. 373–389. Springer; Berlin (2007). International Conference COSIT

Roberts, F.S.: Tolerance geometry. Notre Dame Journal of Formal Logic **14**(1), 68–76 (1973)

Röhrig, R.: Repräsentation und Verarbeitung von qualitativem Orientierungswissen. Ph.D. thesis, University of Hamburg (1998)

Rummery, G.A., Niranjan, M.: On-line Q-learning using connectionist systems. Tech. Rep. CUED/F-INFENG/TR 166, Cambridge University Engineering Department (1994)

Russell, S.J., Norvig, P.: Artificial Intelligence: A Modern Approach (2nd Edition). Prentice Hall Series in Artificial Intelligence. Prentice Hall (2003)

Schiffer, S., Ferrein, A., Lakemeyer, G.: Qualitative world models for soccer robots. In: Proceedings of the Workshop on Qualitative Constraint Calculi: Application and Integration at KI 2006, pp. 3–14. Bremen, Germany (2006)

Schlieder, C.: Anordnung und Sichtbarkeit – eine Charakterisierung unvollständigen räumlichen Wissens. Ph.D. thesis, Computer Science Department, University of Hamburg, Germany (1991)

Schlieder, C.: Representing visible locations for qualitative navigation. In: Proceedings of the Workshop on Qualitative Reasoning and Decision Technologies (QUARDET), pp. 523–532. Barcelona, Spain (1993)

Sierra, C., de Mántaras, R.L., Busquets, D.: Multiagent bidding system for robot qualitative navigation. In: Castelfranchi, C., Lespérance, Y. (eds.) Intelligent Agents VII. Agent Theories Architectures and Languages, 7th International Workshop, ATAL 2000, Boston, MA, USA, July 7–9, 2000, Proceedings, *Lecture Notes in Computer Science*, vol. 1986, pp. 198–212. Springer (2001)

Singh, S.P., Jaakkola, T., Jordan, M.I.: Reinforcement learning with soft state aggregation. In: Tesauro, G., Touretzky, D., Leen, T. (eds.) Advances in Neural Information Processing Systems: Proceedings of the 1994 Conference, vol. 7. MIT Press, Cambridge, MA (1995)

Singh, S.P., Sutton, R.S.: Reinforcement learning with replacing eligibility traces. Machine Learning **22**(1), 123–158 (1996)

Sloman, A.: Why we need many knowledge representation formalisms. In: Bramer, M.A. (ed.) Research and Development in Expert Systems – Proceedings of the BCS Expert Systems Conference, pp. 163–183. Springer (1985)

Smart, W.D., Kaelbling, L.P.: Practical reinforcement learning in continuous spaces. In: Proceedings of the Seventeenth International Conference on Machine Learning (ICML), pp. 903–910. Morgan Kaufmann Publishers Inc., San Francisco, CA, USA (2000)

Soni, V., Singh, S.: Using homomorphisms to transfer options across continuous reinforcement learning domains. In: Proceedings of the National Conference on Artificial Intelligence (AAAI), pp. 494–499. Boston, MA (2006)

Sorrows, M.E., Hirtle, S.C.: The nature of landmarks for real and electronic spaces. In: Freksa, C., Mark, D.M. (eds.) Spatial Information Theory: Cognitive and Computational Foundations of Geographic Information Science. Conference on Spatial Information Theory (COSIT), pp. 37–50. Springer, Berlin (1999)

Stell, J.G., Worboys, M.F.: Generalizing graphs using amalgamation and selection. In: Güting, R.H., Papadias, D., Lochovsky, F. (eds.) Advances in Spatial Databases: Proceedings of the 6th International Symposium on Spatial Databases (SSD), *Lecture Notes in Computer Science*, vol. 1651, pp. 19–32. Springer-Verlag, Berlin Heidelberg (1999)

Sugiyama, M., Hachiya, H., Towell, C., Vijayakumar, S.: Value function approximation on non-linear manifolds for robot motor control. In: Proceedings of the IEEE Conference on Robots and Automation (ICRA) (2007)

Sutton, R.S.: Learning to predict by the method of temporal differences. Machine Learning **3**, 9–44 (1988)

Sutton, R.S.: Generalization in reinforcement learning: Successful examples using sparse tile coding. In: Touretzky, D.S., Mozer, M.C., Hasselmo, M.E. (eds.) Advances in Neural Information Processing Systems: Proceedings of the 1995 Conference, vol. 8, pp. 1038–1044. MIT Press, Cambridge, MA (1996)

Sutton, R.S., Barto, A.G.: Reinforcement learning: an introduction. Adaptive Computation and Machine Learning. MIT Press, Cambridge, MA (1998)

Sutton, R.S., Maei, H.R., Precup, D., Bhatnagar, S., Silver, D., Szepesvári, C., Wiewiora, E.W.: Fast gradient descent methods for temporal-difference learning with linear function approximation. In: Proceedings of the Twenty Sixth International Conference on Machine Learning (ICML). Montreal, Canada (2009a)

Sutton, R.S., Precup, D., Singh, S.: Intra-option learning about temporally abstract actions. In: In Proceedings of the Fifteenth International Conference on Machine Learning (ICML), pp. 118–126. Morgan Kaufmann (1998)

Sutton, R.S., Precup, D., Singh, S.: Between MDPs and semi-MDPs: A framework for temporal abstraction in reinforcement learning. Artificial Intelligence **112**(1-2), 181–211 (1999)

Sutton, R.S., Szepesvári, C., Maei, H.R.: A convergent $O(n)$ algorithm for off-policy temporal difference learning with linear function approximation. In: Koller, D., Schuurmans, D., Bengio, Y., Bottou, L. (eds.) Advances in Neural Information Processing Systems 21: Procedings of the 2008 Conference, pp. 1609–1616. MIT Press (2009b)

Talmy, L.: How language structures space. In: Pick Jr., H.L., Acredolo, L.P. (eds.) Spatial Orientation: Theory, Research, and Application, pp. 225–282. Plenum, New York (1983)

Taylor, M.E., Stone, P.: Behavior transfer for value-function-learning-based reinforcement learning. In: Proceedings of the Fifth International Joint Conference on Autonomous Agents and Multiagent Systems (AAMAS), pp. 53–59. Utrecht, Netherlands (2005)

Taylor, M.E., Stone, P.: Cross-domain transfer for reinforcement learning. In: Proceedings of the Twenty Fourth International Conference on Machine Learning (ICML). Corvallis, Oregon (2007)

Tesauro, G.: Practical issues in temporal difference learning. Machine Learning **8**, 257–277 (1992)

Theocharous, G., Mahadevan, S., Kaelbling, L.P.: Spatial and temporal abstraction in POMDPs applied to robot navigation. Computer Science and Artificial Intelligence Laboratory Technical Report MIT-CSAIL-TR-2005-058, Massachusetts Institute of Technology (MIT), Cambridge, MA, USA (2005)

Thrun, S.: The role of exploration in learning control. In: White, D.A., Sofge, D.A. (eds.) Handbook of Intelligent Control: Neural, Fuzzy and Adaptive Approaches. Van Nostrand Reinhold, Florence, Kentucky 41022 (1992)

Thrun, S.: Is learning the n-th thing any easier than learning the first? In: Touretzky, D.S., Mozer, M.C., Hasselmo, M.E. (eds.) Advances in Neural Information Processing Systems: Proceedings of the 1995 Conference, vol. 8, pp. 640–646. MIT Press (1996)

Thrun, S., Schwartz, A.: Finding structure in reinforcement learning. In: Tesauro, G., Touretzky, D., Leen, T. (eds.) Advances in Neural Information Processing Systems: Proceedings of the 1994 Conference, vol. 7. MIT Press, Cambridge, MA (1995)

Timmer, S., Riedmiller, M.: Abstract state spaces with history. In: Proceedings of the Annual Meeting of the North American Fuzzy Information Processing Society (NAFIPS), pp. 661–666 (2006)

Timmer, S., Riedmiller, M.: Fitted Q-iteration with CMACs. In: Proceedings of the IEEE International Symposium on Approximate Dynamic Programming and Reinforcement Learning, pp. 1–8. Honolulu, HI (2007)

Torrey, L., Shavlik, J., Walker, T., Maclin, R.: Using advice to transfer knowledge acquired in one reinforcement learning task to another. In: Proceedings of the Sixteenth European Conference on Machine Learning (ECML), pp. 412–424. Bonn, Germany (2005)

Torrey, L., Shavlik, J., Walker, T., Maclin, R.: Skill acquisition via transfer learning and advice taking. In: Proceedings of the Seventeenth European Conference on Machine Learning (ECML), pp. 425–436. Berlin, Germany (2006)

Tsitsiklis, J.N.: Asynchronous stochastic approximation and Q-learning. Machine Learning **16**, 185–202 (1994)

Tsitsiklis, J.N., Van Roy, B.: An analysis of temporal-difference learning with function approximation. IEEE Transactions on Automatic Control **42**(5), 674–690 (1997)

Uther, W.T.B., Veloso, M.M.: Tree based discretization for continuous state space reinforcement learning. In: Proceedings of the National Conference on Artificial Intelligence (AAAI), pp. 769–775. Madison, WI (1998)

Uther, W.T.B., Veloso, M.M.: TTree: Tree-based state generalization with temporally abstract actions. In: Alonso, E., Kudenko, D., Kazakov, D. (eds.) Adaptive Agents and Multi-Agent Systems: Adaptation and Multi-Agent Learning, *Lecture Notes in Artificial Intelligence*, vol. 2636, pp. 260–290. Springer-Verlag Berlin Heidelberg (2003)

Vollbrecht, H.: Hierarchic function approximation in kd-Q-learning. In: Proceedings of the Fourth International Conference on Knowledge-Based Intelligent Engineering Systems & Allied Technologies. Brighton, UK (2000)

Wagner, T.: Qualitative sicht-basierte Navigation in unstrukturierten Umgebungen. Ph.D. thesis, Fachbereich 3 (Mathematik und Informatik), Universität Bremen (2006)

Wagner, T., Huebner, K.: An egocentric qualitative spatial knowledge representation based on ordering information for physical robot navigation. In: Nardi, D., Riedmiller, M., Sammut, C. (eds.) RoboCup 2004: Robot Soccer World Cup VIII, *Lecture Notes in Artificial Intelligence*, vol. 3276, pp. 134–149. Springer, Heidelberg (2005)

Wallgrün, J.O., Frommberger, L., Wolter, D., Dylla, F., Freksa, C.: Qualitative spatial representation and reasoning in the SparQ-toolbox. In: Spatial Cognition V: Reasoning, Action, Interaction: International Conference Spatial Cognition 2006. Bremen, Germany (2007)

Watkins, C.: Learning from delayed rewards. Ph.D. thesis, Cambridge University (1989)

Watkins, C., Dayan, P.: Q-learning. Machine Learning **8**, 279–292 (1992)

Weaver, S., Baird, L., Polycarpou, M.: An analytical framework for local feedforward networks. IEEE Transactions on Neural Networks **9**(3), 473–482 (1998)

Webber, B.L., Nilsson, N.J. (eds.): Readings in Artificial Intelligence. Tioga Publishing Company, Palo Alto, CA (1981)

Whitehead, S.D.: Reinforcement learning for the adaptive control of perception and action. Ph.D. thesis, University of Rochester, Department of Computer Science, Rochester, New York (1992)

Whitehead, S.D., Ballard, D.H.: Learning to perceive and act by trial and error. Machine Learning **7**(1), 45–83 (1991)

Whiteson, S., Stone, P.: Evolutionary function approximation for reinforcement learning. Journal of Machine Learning Research **7** (2006)

Wilson, S.W.: Explore/exploit strategies in autonomy. In: Maes, P., Mataric, M., Pollac, J., Meyer, J.A., Wilson, S. (eds.) From Animals to Animats 4: Proceedings of the Fourth International Conference on Simulation of Adaptive Behavior, pp. 325–332. MIT Press, Cambridge, MA (1996)

Wolfe, A.P., Barto, A.G.: Decision tree methods for finding reusable MDP homomorphisms. In: Proceedings of the National Conference on Artificial Intelligence (AAAI), pp. 530–535. Boston, MA (2006a)

Wolfe, A.P., Barto, A.G.: Defining object types and options using MDP homomorphisms. In: Proceedings of the ICML Workshop on Structural Knowledge Transfer for Machine Learning. Pittsburgh, PA, USA (2006b)

Wolter, D.: SparQ—a spatial reasoning toolbox. In: Proceedings of AAAI Spring Symposium on Benchmarking of Qualitative Spatial and Temporal Reasoning Systems (2009)

Zadeh, L.A.: Fuzzy sets and information granularity. In: Gupta, M.M., Ragade, R.K., Yager, R.R. (eds.) Advances in Fuzzy Set Theory and Applications, pp. 3–18. North Holland Publishing Company (1979)

Zhao, G., Tatsumi, S., Sun, R.: A heuristic Q-learning architecture for fully exploring a world and deriving an optimal policy by model-based planning. In: Proceedings of the IEEE Conference on Robots and Automation (ICRA), pp. 2078–2083. Detroit, MI (1999)

Index

T

U

V